The Chanterelle Book

The Chanterelle Book

by

Olle Persson

Illustrations by

Bo Mossberg

Ten Speed Press
Berkeley, California

For further information, write to:

Ten Speed Press
P.O. Box 7123
Berkeley, California 94707

Distributed in Australia by Simon & Schuster Australia, in Canada by Publishers Group West, in New Zealand by Tandem Press, in South Africa by Real Books, and in the United Kingdom and Europe by Airlift Books.

Cover by Catherine Jacobes

Bibliography is on file with publisher.

Printed in Hong Kong
First Printing, 1997
1 2 3 4 5 — 01 00 99 98 97

Library of Congress Cataloging-in-Publication Data

Persson, Olle, 1928–
[Kantareller. Swedish]
Chanterelles and related species of
Europe and North America / by Olle Persson.
p. cm.
Includes bibliographical references and index.
ISBN 0–89815–947–4
1. Cantharellus. I. Title.
QK629.C3P4713 1997
579.5'97—dc21 97–15957
CIP

Table of Contents

Det är kantarelliskt gulrusigt
att leva någon dag
som i längst förgångna
spånkorghöstar.
Ändå är de här.
En återkomstens livslust
drar uppsluppet genom vår skog.
Det jordtillvända solskenet
ger oss plötsligt ärende bort
till vårt eget älskade land.
Vi går ut på gångstigsforskning.

It is a chanterellishly yellow high
to live for a day as in
a chip-basket autumn
of long ago.
Still they come.
The excitement of returning
is trekking happily through our forest.
Sunshine beating down on the Earth
suddenly gives us reason to visit
our own beloved land.
We step out to research the trail.

Harry Martinsson, from *De tusen dikternas bok*
(Book of a Thousand Poems), 1986

Translated by Dag Forssell

Preface

It is not uncommon to hear mushroom pickers say, "I only pick chanterelles." The purpose of this book is to discuss chanterelles and some related mushrooms (so-called chanterelloid mushrooms) and thus provide the reader with a deeper insight into a biologically interesting mushroom group with a rich cultural history. Hopefully, it will also awaken the reader's interest in mycology in general. Mushrooms have lately been the focus of much interest. One reason is the increase in environmental consciousness; however, this has not always been the case. In the past, mushrooms were considered "cattle feed" in, for example, U.S.A. and Sweden, but in other parts of the world, mushrooms have been part of our diet for thousands of years.

Carolus Linnaeus wrote in his book *Västgötaresa* (Journey in Västergötland, 1746), "Large numbers of chanterelles, an edible mushroom, grew on Hunneberg. . . ." But it was most likely not prepared in the typical farmers' kitchens. With the spreading influence of French cuisine through Europe, the chanterelle first gained recognition in palace kitchens. During the latter half of the nineteenth century, especially during years of famine, interest in mushrooms increased. Another contributing factor was education: School became compulsory, and the study of mushrooms gained greater exposure. But people did not really become interested in mushrooms until the war years in this century. We then also started to pick *Cantharellus tubaeformis* and *Craterellus cornucopioides*.

In this book, European and North American chanterelles and some related mushrooms are discussed from the multiple perspectives of biology, ecology, geography, culinary science, culture, and linguistics. We have added to our knowledge by means of literary studies and travel, as well as through collaboration with mycologists around the world.

The beauty of mushrooms, both their shapes and their colorings, has also inspired us. Painting a mushroom may seem mundane to an artist, but there is joy to be found in using the tip of the brush to capture all the structures, color nuances, shadows, and light areas of a mushroom. The orange chanterelle's (*Cantharellus friesii*) silky, curved, fine-veined cap forms an exciting miniature landscape. It is marvelous for an artist to reproduce all the hues of an autumn leaf, the redness of a lingonberry, or the metallic shine of a dung beetle.

Our collaboration inspired an idea for the illustrations. We felt it was natural to depict each species using three different picture types. The first is a sketch of the mushroom site seen from above. The second shows various stages of the mushroom's growth with small inlays of specimens that also illustrate the mushroom environment. The compositions on these pages were inspired by the beautiful illustrations by Elias Fries, in *Sveriges ätliga och giftiga svampar* (Edible and Poisonous Mushrooms in Sweden). The third picture type is of the entire environment. Although a certain species may be found in many different environments, we have chosen *one* typical setting. All paintings have been done using living models. Pictures have either been preceded by careful sketches or painted outside in nature.

Working on the book has been very stimulating. Without all the contacts we have established both in Sweden and abroad, we would not have been able to bring our ideas to life. There is not enough space here to mention all those who have helped us, but we would like to especially thank a few of them. Ronald H. Peterson, from Knoxville, Tennessee, in the U.S., one of the foremost experts on chanterelles, awoke our interest in them. Without the travels he and Olle Persson embarked upon in the 1970s in the United States and Sweden, the book would never have materialized. Discussions with Eric Danell, of Uppsala University, and Nils Lundqvist and Åke Strid, of the Swedish Natural History Museum in Stockholm, solved many problems. We also received some much-appreciated help from Mats Rydén, of Uppsala. His extensive knowledge of linguistic problems and the naming of plants has enriched the book.

From among those who have helped us with mushroom material and information, we also would like to thank Elsa Bohus-Jensen, Pelle Holmberg, Raoul Iseborg, Stig Jacobsson, Siw Muskos, and Staffan Åström, from Sweden; Henning Knudsen and Jan Vesterholt, from Denmark; Mauri Korhonen and Ilkka Kytövuori, from Finland; Gro Gulden, Claus Høiland, Leif Ryvarden, Sigmund Sivertsen, Ola Skifte, and Jens Stordal, of Norway. We received both field assistance and help with the material from others throughout the rest of Europe. Thomas Laessøe and David Pegler helped us when we went through the Kew Herbarium in London, and Jakob Elmer and Franco Patané helped us with Swiss material. Jörg

Thien and Hermine Wechner helped us to acquire the material from Austria, which made the book almost complete in regard to Europe. Elizabeth and Hendricks Kern in Seattle helped us with material for the pictures of the magnificent chanterelles from the west coast of North America.

We are grateful not only to the above-mentioned people, but also to all of our friends in mycological associations both here and far away—no one has been forgotten.

And finally, we would like to thank our publisher, who invested in our idea and gave us much support.

Olle Persson *Bo Mossberg*

The Swedish edition of this book has been appreciated by foreign mycologists and mushroom lovers alike. However, they urged Olle Persson and Bo Mossberg to investigate the possibilities for a translation into English. This American edition is a translation of the Swedish book. However, the translated text also includes more data of interest to American mycologists, and some details from Sweden have been omitted. The authors and I decided to maintain the original systematical concept of the Swedish edition, mainly based on Kühner & Romagnesi (1953). However, modern data derived from DNA analyses by, for example, Toby Feibelman imply that there are only two genera in the family Cantharellaceae: *Cantharellus* and *Craterellus*. *Craterellus* should therefore include species with thin flesh such as *C. tubaeformis* and *C. undulatus*. It should also be emphasized that the European yellow chanterelle *Cantharellus cibarius* is often mixed up with exotic species. The commonly found and highly appreciated Pacific Golden Chanterelle from the North American Pacific Northwest is not *Cantharellus cibarius*, but *Cantharellus formosus*. To avoid confusion in scientific results and food quality, it is of great importance to ecologists and traders to put the correct label on the organism treated. Finally, a paragraph on a new cultivation technique has been included, giving some background to successful fruitbody formation in a greenhouse in 1996. I wish to acknowledge some of my American colleagues who generously encouraged my U.S. research and field studies: David Arora, Francisco Camacho, Toby Feibelman, Aaron Liston, Randolph Molina, Lorelei Norvell, Ronald Petersen, David Pilz, Scott Redhead, and James Trappe.

Eric Danell, May 1997

What are Chanterelles?

If we ask a mushroom picker to define chanterelles, he or she may answer, "yellow mushrooms." But what is a mushroom, and how do chanterelles differ from other mushrooms? Let us first discuss the historical background of mushrooms.

When we look back in time, we realize that the interest of humans in the living things surrounding them used to be primarily linked to utility—and danger. Six thousand years ago, the Sumerians of the Tigris and Euphrates Rivers' delta were already aware of the medicinal properties of plants, and cultivated them as well. Humans have used domesticated animals since ancient times; hunting and fighting wild animals is depicted on cave paintings more than 30,000 years old. Similarly, humans learned early on to distinguish between usefulness and danger with plants. In books about herbs written in the sixteenth century, useful plants were added as they were adopted.

When, then, did we start to pay attention to mushrooms? We know that mushrooms were first used for medicinal purposes a long time ago: The "Iceman," whose body was found in the Alps in 1991, had been frozen for more than 5,000 years, and pieces of an antibiotic polypore were found on him. The ancient Greeks and Romans had names for a number of mushrooms that were used both in the kitchen and in medicine. But their origin was shrouded in mystery. Herbalists who wrote books on them still did not know much of the nature of mushrooms. The German herbalist Jerome Bock (Hieronymus Tragus) wrote in 1552, "They are neither medicinal herbs, roots, flowers nor seeds, but rather a manifestation of excess moisture from soil, trees, rotten wood and other rotten things. It is a known fact that especially those which are eaten most grow best in humid, thunderous weather." This was before the invention of the microscope and the discovery of spores, the "seeds" of mushrooms. In his binomial nomenclature of plants, Linnaeus classified mushrooms, *fungi*, in the same group as ferns, mosses, and algae, or as *Cryptogamia* (the Greek word *kryptos* means concealed, and *gamos* means wedding). He wrote, "The classification of mushrooms is still chaotic for the botanist." It was Pliny (see page 119) who coined the term "fungi" for all mushrooms except Caesar's Mushroom (an edible fly agaric), which was called Boletus.

The persistent conception that everything living belongs either to the animal kingdom or to the plant kingdom has been abandoned. The current view is that there are five different kingdoms, which include two groups—organisms without cell nuclei, or Procaryota (the Greek word *pro* means before, and *karyon* means core), and those with one, Eucaryota (the Greek word *eu* means real). Bacteria including blue-green algae (cyanobacteria) belong to the first group, while animals, plants, and fungi belong to the other. In Procaryota the genetic material is linked to rings of free DNA in the cells, while it is contained in the cell nucleus in Eucaryota. This is significant for the definition of mushrooms and for the distinction between chanterelles and other mushrooms.

One definition of chanterelles follows:

Basidiomycetes with a cap and stalk, and a hymenium which cannot be removed from the cap and which is spread across the surface of blunt folds or ridges, or an almost even surface on the underside of the cap. The basidia are often stichobasidia.

According to one modern definition, mushrooms are organisms without chlorophyll, the cells of which are nucleated and are able to absorb nutrition through the outer layer of their cell walls. This definition requires the use of more advanced microscope equipment and methods of chemical analysis, something to which mycologists of old did not have access. Among the large mushrooms, there are two primary groups: sac fungi, or Ascomycetes, and Basidiomycetes. The spores of Basidiomycetes are often formed on club-shaped outgrowths called *basidia*. The chanterelle is a Basidiomycete. The basidia are formed on a skinlike surface called *hymenium* (the Greek word *hymen* means skin). This can develop on the surface of lamellae as is the case with a large group of mushrooms (e.g., agarics), but also in tubes as in Boletes and Polypores, on the surface of spines as with hedgehog mushrooms, and on a clublike or branched surface as with Clavaria.

One characteristic that has received more attention today is the way the hymenium grows, for example, whether the basidia develop simultaneously across the entire hymenium or grow successively. In chanterelles they grow successively and are stored on top of each other. This is not the case in, for example, *Agaricus bisporus*. This results in the thick, rounded lamellalike ridges in chanterelles. They also are extensively branched. In some species (e.g., *Craterellus cornucopioides*) they are almost totally flattened.

Another important distinction used today is how the cell nucleus is divided in the basidium.

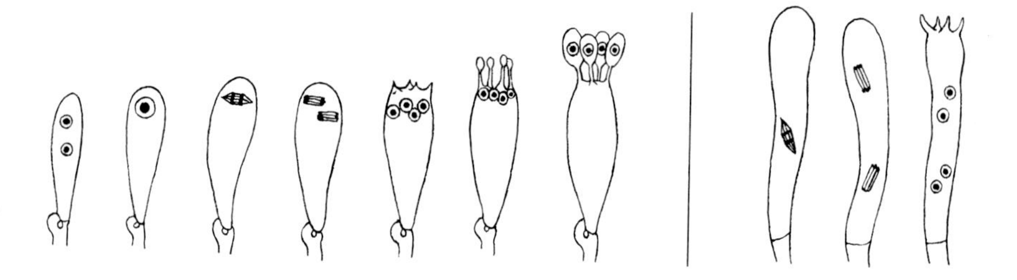

Cell division in chiastobasidia in 7 phases (left) and in stichobasidia in 3 phases (right), corresponding to phases 3, 4, 5 for chiastobasidia

Chanterelles have elongated, almost cylindrical basidia, and the division occurs lengthwise. These are referred to as *stichobasidia* (the Greek word *stichos* means row). In agarics, and many other fungi, the division occurs at a right angle to the longitude of the basidium. These are referred to as *chiastobasidia* (the Greek word *chiazein* means mark crosswise) and are often thicker and more clublike.

A third characteristic is if the *hyphae* have "clamp connections," half-circle-shaped protrusions on the wall between two adjacent cells. This characteristic is often a distinguishing mark on a genus level. These clamp connections are common among Basidiomycetes. We will discuss this in more detail when we talk about the different species on page 32.

Some mushrooms which earlier were included among the chanterelles, based only on their shape, are now referred to as "chanterelloid mushrooms" (see page 32).

Mushrooms in History

The oldest known picture of mushrooms is a fresco painting that was buried in lava when Vesuvius erupted in 79 AD. It has been interpreted as depicting a milk cap in the *Lactarius deliciosus* group, a mushroom mentioned by Classical-era authors. The first illustrations of mushrooms were woodcuts and can be found in *Ortus sanitatis* from 1491, and in *The grete herball* from 1526. These pictures are difficult to identify. In an herbal book from 1560, the Italian Mattiolo included better illustrations. He depicted species that had been mentioned by the Greek Dioscorides during the first century, among other mushrooms, in addition to truffles and the medicinal polypore, *Fomes officinalis*, also called Agaricon. The polypore is clearly shown growing on a larch tree.

To the best of our knowledge, the first picture of a chanterelle is from 1581. It was included in the book *Kruydtboeck* (the Dutch word *kruydt* means herb), by Mathias de l'Obel (Lobelius). This illustration is reproduced below.

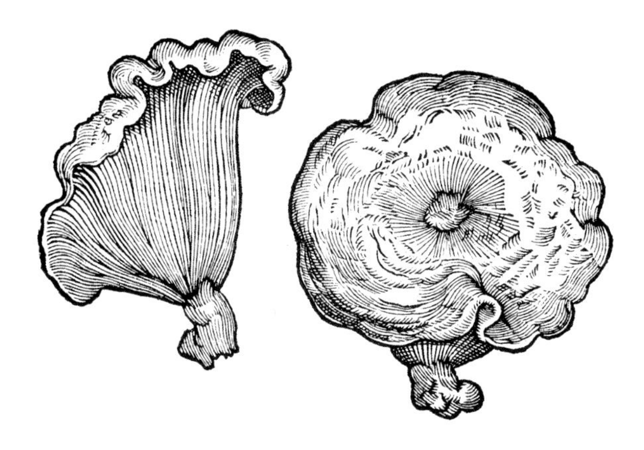

In 1601, the first flora book containing high-quality illustrations of mushrooms was published. It was Carolus Clusius's *Rariorum plantarum historia,* in which mushrooms are discussed in a special appendix under the title *Fungorum in Pannoniis observatorum brevis historia* ("Fungorum historia"). Clusius was the first to write a natural scientific mushroom monograph. He also realized the importance of reproducing mushrooms in color. Clusius is considered to be one of history's foremost botanists, and was active in many areas in Europe.

After arriving in Vienna in 1573, Clusius first worked for the Emperor Maximilian II and headed the Vienna Botanical Garden. After the death of the Emperor, he lost his position but received financial support from the Hungarian Count Boldizsár Batthyány. Together with the court priest, Istvan Bythe, they made numerous excursions to the old Roman province Pannonia, which included parts of Austria, Hungary, and Yugoslavia. These studies resulted in the book "Fungorum historia." Batthyány and Bythe helped Clusius with, among other things, recording popular names of mushrooms in Hungarian, German, and Slavic languages (see pages 16, 30). He also received help in painting the mushrooms in color from a French watercolor painter living in Vienna. We will discuss Clusius's pictures in more detail on page 16.

Clusius had a number of successors. Many of them used his and Lobelius's illustrations. Frans van Steerbeeck, for example, used Clusius's *Codex* (see page 16) as a basis for his work, *Theatrum fungorum,* which was published in 1675.

During the eighteenth century a new technique, copperplate engraving, was introduced. In 1727, the illustrator Claude Aubriet used the new technique in Sébastien Vaillant's book, *Botanicon parisiense,* to draw the chanterelle and *Craterellus cornucopioides* or "Horn of Plenty" (see next page). Other important works with hand-colored copperplate engravings were published during the 1700s. The German Jacob Christian Schaeffer (1718–1790) and the Frenchman Jean-Baptiste François Bulliard (1752–1793) produced a large volume of posters that provided the basis for the work of mycologists in the nineteenth century. Schaeffer provided illustrations of *Cantharellus cibarius, Cantharellus aurora, Craterellus cornucopioides,* and *Gomphus clavatus,* while in his works, Bulliard had illustrations of *Cantharellus cibarius, Cantharellus tubaeformis, Craterellus cinereus, Cantharellus aurora,* and *Craterellus cornucopioides.* The great mycologists of the nineteenth century referred to these illustrated works in their descriptions.

Chanterelle (Cantharellus cibarius) *and Horn of Plenty* (Craterellus cornucopioides) *from* Botanicon parisiense

During the nineteenth century, the foundation of modern mycology was laid by the Dutchman Christiaan Hendrik Persoon and the Swede Elias Fries.

Persoon reproduced some of his species in *Fungorum minus cognitorum*, which contains illustrations of *Craterellus cinereus*. Fries's important work from the first half of the nineteenth century did not have any illustrations. On the other hand, he published two monumental illustrated works during the latter half of the nineteenth century. We will mention more about these when we show some pictures from one of these works.

From 1831 to 1846, the physician and mycologist Julius Vincenz von Krombholz published an illustrated book in Prague with a very insightful text. He is likely the first to have addressed the subjects of poisonous mushrooms and the nutritional value of mushrooms in more detail, having tried many mushrooms himself. Krombholz and his illustrations are discussed in more detail on page 20.

Chanterelles According to Clusius

Early on, Clusius became aware that black-and-white illustrations were not adequate when describing such a uniform group as agarics. With Batthyány's financial support, he made color illustrations of eighty-seven mushrooms. These illustrations were collected in Clusius's *Codex*, and, because they were unsigned, their provenance was a mystery for a long time. The pictures disappeared before "Fungorum historia" was printed and were lost until the seventeenth century. Instead, simpler woodcuts were used in "Fungorum historia." One of these, depicting a chanterelle, *Cantharellus cibarius*, is shown below.

Clusius's original illustrations then ended up in Leiden. The name of the artist was discovered in the beginning of the twentieth century. Art historians were able to find proof that the French artist most likely was Clusius's nephew who had spent time in Vienna as a portrait painter.

The water paintings in the *Codex* include handwritten popular names of the mushrooms.

GENUS XIIII.

XIIII. *gen. ęscul. fung.* Nivvl gõba. Hasen ör-lein.

EODEM, quo ſuperioris generis ſpecies, tẽpore in ſilvis oritur *Decimum-quartum* eſculentorum fungorum genus, quod Vngari *Niwl gomba*, hoc eſt, Leporinum fungum: Germani Haſen örlein Leporinam auriculam, appellant. Videtur autem in duas ſpecies poſſe diſtingui: niſi quis forſitan ætate duntaxat differre putat.

I. *Prioris* ſpeciei forma anguloſa eſt, magis tamen in orbẽ circinata, unciam, aut paulo amplius lata, brevi craſsóq; pediculo ſubnixa, ſupernâ parte extuberans, pallidi coloris, quem flavæ aliquot maculæ varium faciunt: inferna pars flaveſcit, fuſciſque ſtrijs exarata eſt.

XIIII. Generis eſculentorum Fungorum ſecunda ſpecies.

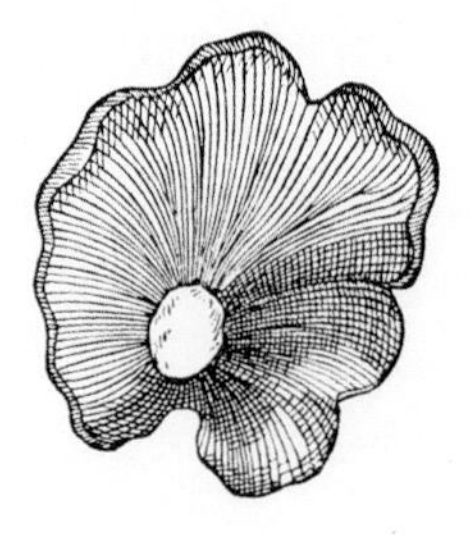

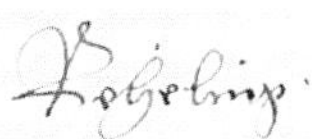

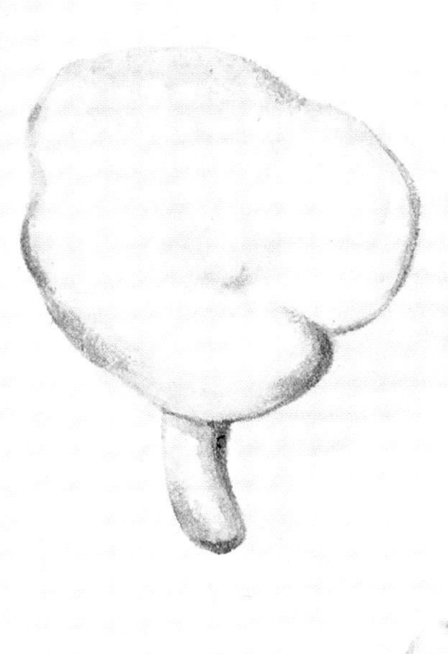

Eighteenth-Century Illustrations

Jacob Schaeffer's work, *Fungorum qui in Bavaria et Palatinatu circa Ratisbonam nascentur icones I–IV*, is one of the most important mushroom works written in the eighteenth century (below and top right). At the end of the century, the Dutchman Persoon published the work, *Icones et descriptiones fungorum minus cognitorum. Craterellus cinereus* is shown at the bottom of the next page.

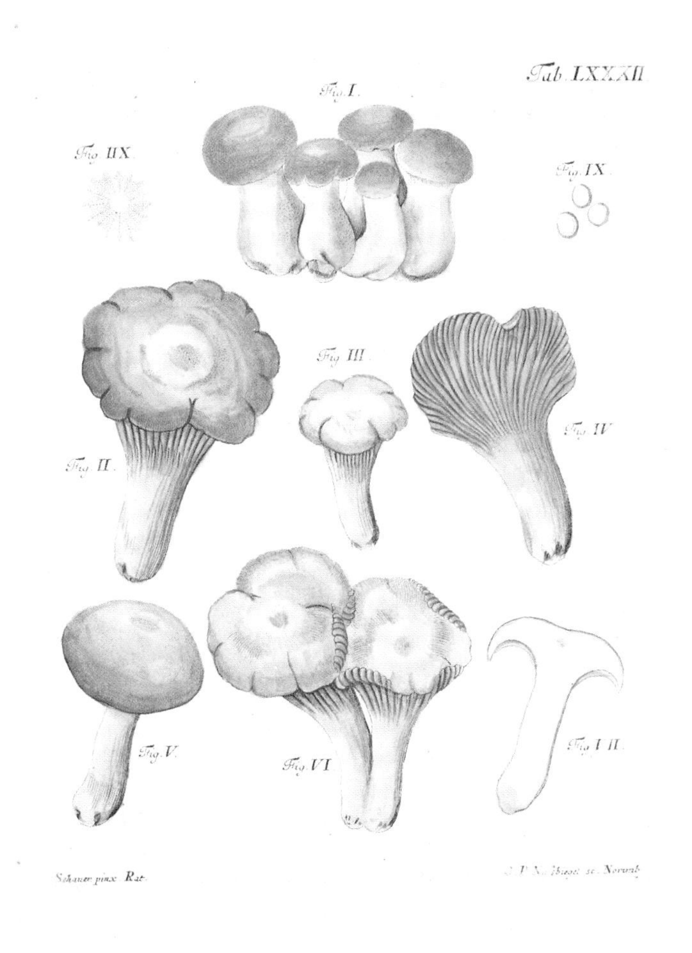
Tab. LXXXII
Fig. I.
Fig. IIX.
Fig. IX.
Fig. III.
Fig. IV.
Fig. II.
Fig. V.
Fig. VI.
Fig. VII.

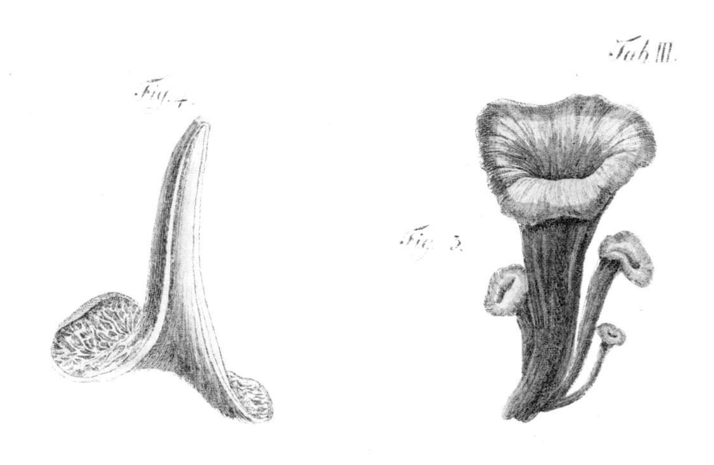
Tab. III.
Fig. 3.

Chanterelles According to Krombholz

Krombholz was active in Prague as a physician and as a professor of pharmacology. He found that textbooks did not provide the necessary information regarding useful and hazardous foods, and that it was his obligation to remedy this, especially where mushrooms were concerned. He therefore decided to write about them, publishing his writings in segments. The first part was published in 1831, and the last one in 1846, just after Krombholz's death. This work was given the comprehensive title, *Naturgetreue Abbildungen und Beschreibungen der essbaren, schädlichen und verdächtigen Schwämme* (Accurate Depictions and Descriptions of Edible, Dangerous, and Suspected Mushrooms, shortened to "Schwämme").

In the preface, which was included in the first booklet, Krombholz noted that he intended to thoroughly discuss terminology and nomenclature both in Latin and German. In addition, he would include names in German and Bohemian dialects, and, if possible, in foreign languages. As far as the illustrations were concerned, he wrote that he would depict species belonging to the same family adjacent to each other, so that it would be easy to distinguish between edible and poisonous mushrooms. The illustrations were rendered accurately under his supervision by skilled botanical illustrators, as lithographs and in color. The next page shows a number of chanterelles; the species depicted are

1-11 Chanterelle, *Cantharellus cibarius*
12 Gray Chanterelle, *Cantharellus cinereus*
13-17 Pig's Ears, *Cantharellus clavatus*
18 Horn of Plenty, *Cantharellus cornucopioides*

Here we have used the Latin names used by Krombholz, who called them all *Cantharellus*. Krombholz's views on the preparation of edible mushrooms are discussed in the chapter entitled "Cooking with Chanterelles" (see page 88).

Chanterelles According to Fries

The Swedish mycologist Elias Fries's scientific works (*Systema mycologicum*, etc.) are fundamental to the worldwide study of mycology. His books are discussed briefly in connection with the description of the various species of chanterelles. There is also a brief review at the end of this book. Here, we will discuss the background of the Royal Swedish Academy of Science's plates, which were produced under Fries's supervision.

Fries became interested in fungi while in school. During the war with Russia from 1808 to 1809, when the school remained closed, he stayed in the countryside and studied fungi on his own. When he was fourteen years old, he described 300 to 400 species. In 1811, he arrived in Lund to begin his university studies. In the university library he found plate collections, such as *Flora danica*, in which he recognized and identified species that he had himself described. During that time, he also learned about Persoon's work, *Synopsis methodica fungorum,* which he soon memorized from cover to cover, according to his autobiography. During his studies, his appreciation of color illustrations grew. He felt that they were the best substitute for dried fungi in herbaria and fungi preserved in alcohol. But it was not until he had established himself as a professor in Uppsala that he could embark on the work that resulted in the two monumental volumes, *Sveriges ätliga och giftiga svampar* (Edible and Poisonous Mushrooms in Sweden) and *Icones selectae hymenomycetum*.

Beginning in 1844, the Royal Swedish Academy of Science provided financial support for the project of illustrating Swedish mushrooms, especially agarics. This support allowed for the commissioning of artists. Mushrooms were mostly collected around the area of Uppsala, but during the years 1854 and 1855, Fries sent his student M. A. Lindblad, together with an artist, to Femsjö in Småland. In a letter to Lindblad, he gave him detailed instructions of where he could find the different species. He also received a number of sketches from his student and friend, H. von Post, which he later had redrawn. E. Pettersson and P. Åkerlund are the artists most often named on the plates. But an artist always develops. The first il-

SVERIGES

ÄTLIGA OCH GIFTIGA SVAMPAR

tecknade efter naturen

under ledning af

E. FRIES

utgifna af

Kongl. Vetenskaps-Akademien.

STOCKHOLM, 1861.
P. A. NORSTEDT & SÖNER

lustrations were not accepted by Fries. In a letter to Lindblad, he wrote that "*Russula adusta* and *nigricans* must be drawn, as the previous drawing was the first one by Pettersson, and it is useless."

The title page of *Sveriges ätliga och giftiga svampar* (Edible and Poisonous Mushrooms in Sweden) is shown above, and on the following pages are illustrations of the chanterelle (*Cantharellus cibarius*) and *Gomphus clavatus* from the same book.

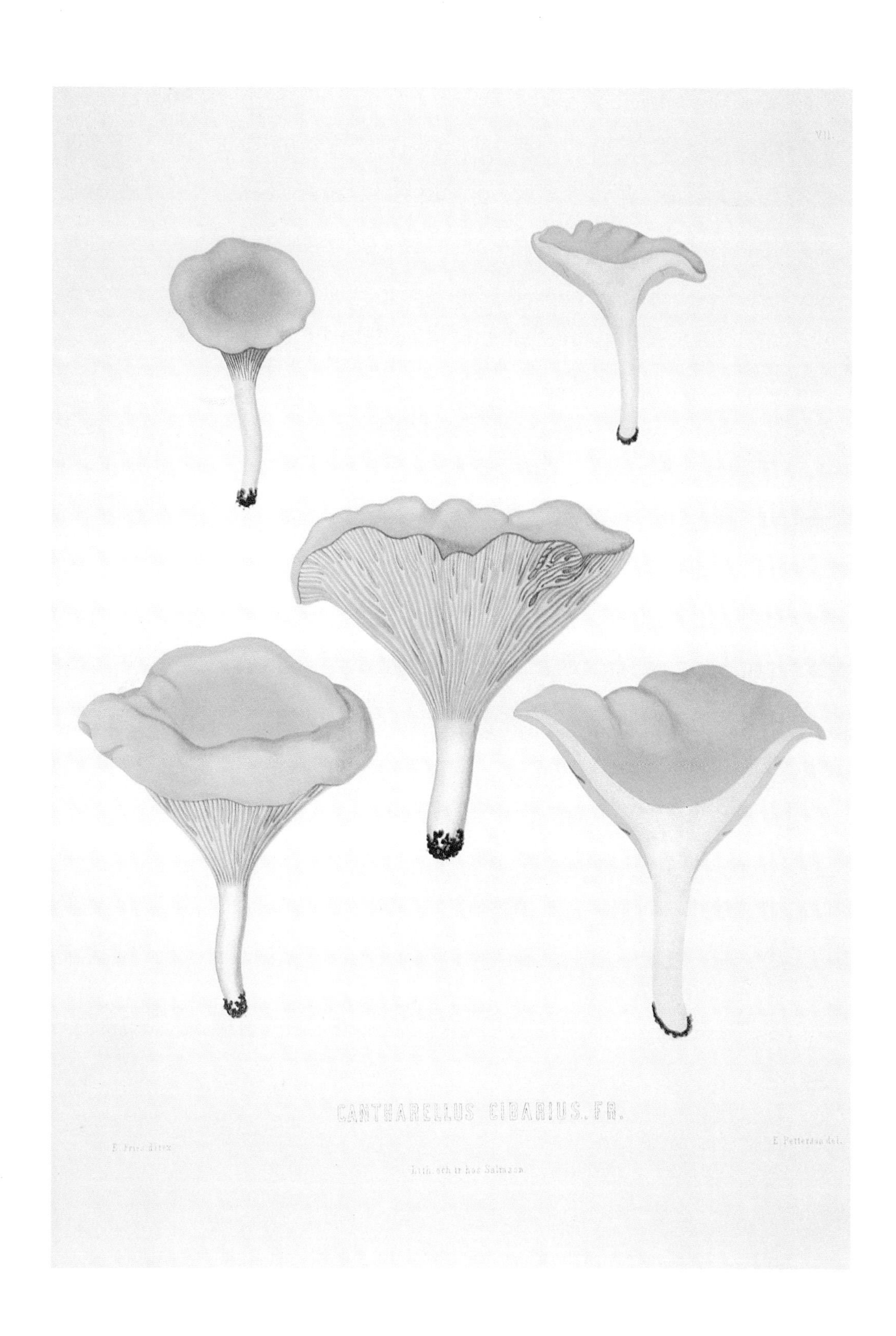
VII.
CANTHARELLUS CIBARIUS. FR.
E. Petterson del.
Lith. och tr hos Salmson

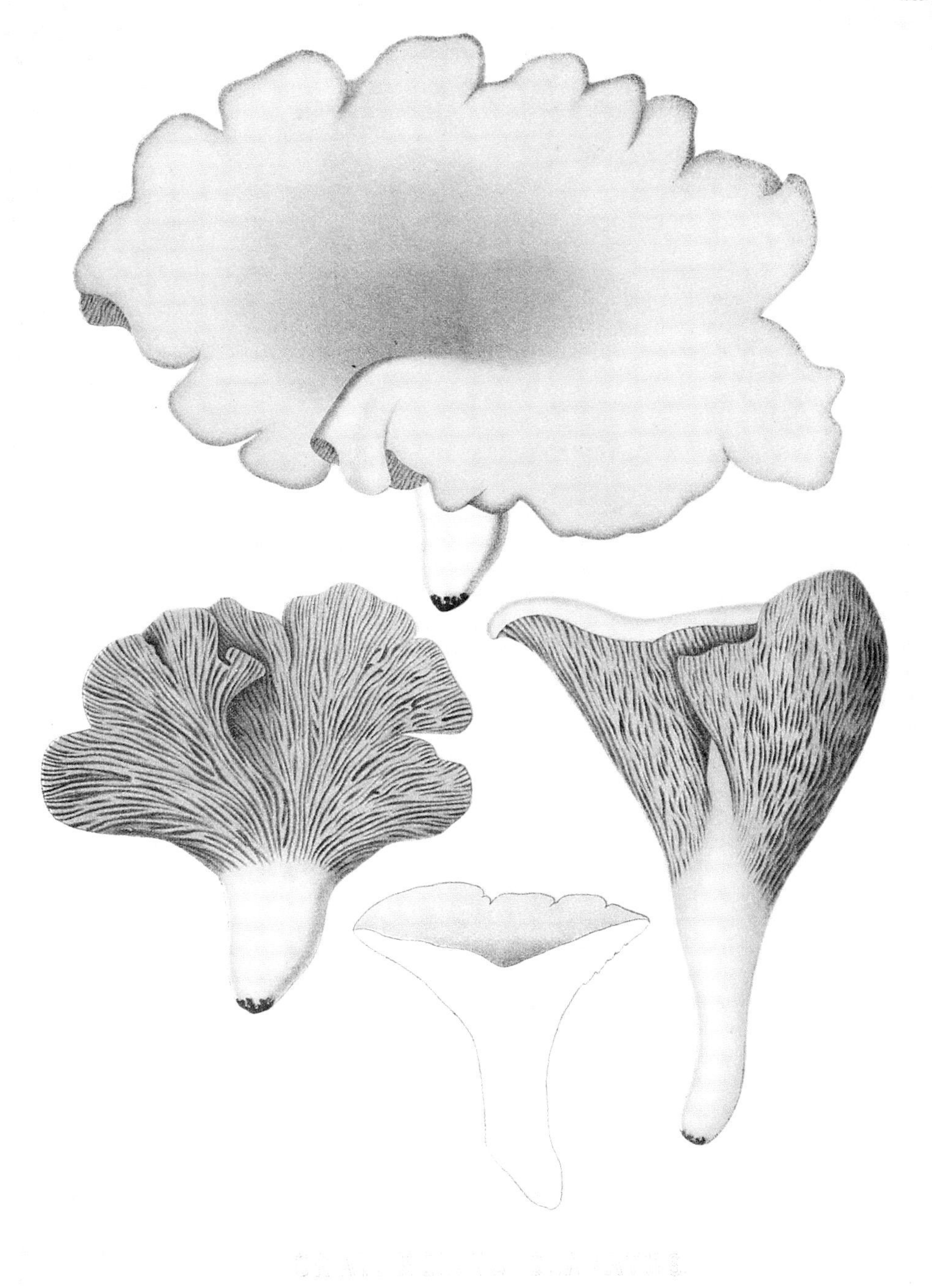

Biology, Ecology, and Geography

It is said that there used to be many more chanterelles in Europe. Is this true, or are such statements based on erroneous information? In older literature, the chanterelle is described as very common. As we know, the mushroom supply varies from year to year. A year with plenty of mushrooms of one species is seldom followed by another equally good year for the same species. In addition, people may base their recollections on childhood mushroom sites that now have totally changed due to land use. Fields, for example, become overgrown again as reforestation begins. In order to make accurate statements, we need a long series of observations from the same place, as the conditions can be quite different in various areas.

However, there is some data indicating that the statement is true. Studies in both the Netherlands and Germany indicate that the supply of chanterelles has decreased. A study of the amount of chanterelles collected in Saarbrücken and offered for sale in the local marketplace during the years 1956 to 1975 showed a drastic drop (see figure below). The chanterelle is now on the German list of endangered species.

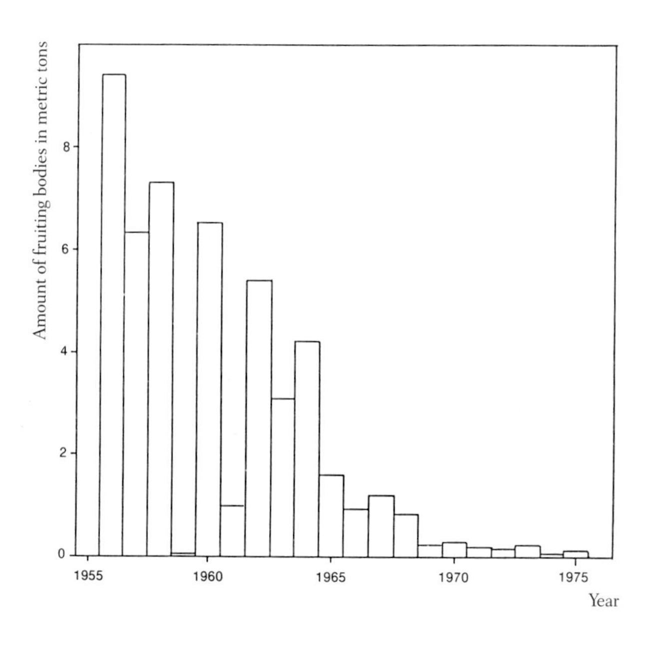

The highest number recorded, more than 9 metric tons in 1956, gives pause if one compares it to the sale of 70 to 90 tons in the Munich market in 1901 (see page 50). The mushroom-growing areas there are different, of course, but the discrepancy is still quite significant.

The reasons for this decrease are being considered. Two reasons under discussion are acidification and increased nitrogen levels in the soil. Another possible factor is changing forestry patterns, including same-age monocultivation. As far as causality is concerned, any hypothesis must be supported by both laboratory and field tests. It has been possible to provide better support for such assumptions, as today chanterelle mycelia can be grown in laboratory settings. Research regarding spore germination by the late Professor Nils Fries, in Uppsala (Sweden), has contributed to understanding these issues. Based on Fries's pioneer work, techniques for routine production of European *Cantharellus cibarius* mycelia, mycorrhiza formation (see below) with pine and spruce, outplanting to greenhouses, and finally fruitbody formation have been developed (Danell 1994 and Danell & Camacho 1997). In essence, the cultivation starts with isolation of mycelia from tissue samples of fruitbodies collected in the field. An important reason why chanterelle cultivation was delayed until 1996 is due to *Pseudomonas* bacteria, which colonize chanterelle fruitbodies in nature. The mycelium is purified by, for example, antibiotics and careful reinoculation. In a sterile environment, the roots of seedlings are colonized by chanterelle mycelia, demanding elevated carbon dioxide and glucose levels. After 8 to 12 weeks, these mycorrhizal seedlings are transferred to pots in a greenhouse, where fruitbodies form after one year. Efforts are being made to adapt this technique to field plantations.

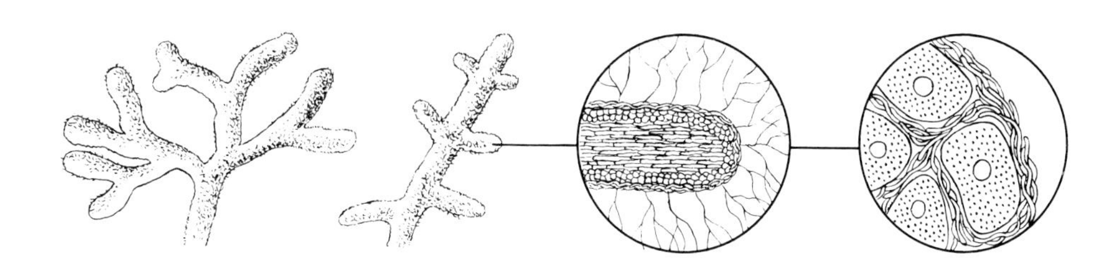

Ectomycorrhizae involving pine (left) and spruce trees, with an enlarged cross section of a root tip, and an additional enlargement of cortex cells surrounded by hyphae

Ectomycorrhiza is a symbiotic relationship between trees and fungi such as chanterelles. The fungus receives sugar from the tree while providing the tree with minerals such as phosphorus. During ectomycorrhizal infection, a mantle of fungal tissue (hyphae) forms around the tree's short roots; the result is a network of hyphae forming around the cortex cells of the roots. In other words, unlike commonly cultivated mushrooms grown on compost or sawdust, chanterelles require the presence of living trees.

The ectomycorrhizal roots of the pine tree become thick and repeatedly bifurcated, while the spruce's roots become more irregular. It is very easy to see the mycorrhiza by removing some moss near a tree. The viability of mycorrhiza is very important in forests. If the trees suffer from acidification, for example, the mycorrhiza is also affected. Acidification may also affect the mycorrhiza and, by extension, the trees.

It is now possible to use DNA technology to determine which species of fungi are linked to trees in forests. A DNA pattern is as revealing as a fingerprint. In addition, the exchange of nutrients between trees and fungi was earlier studied using radioactively marked nutrients.

There are primarily two factors that are of importance for mycorrhizal fungi: the amount of nitrogen in the soil and the supply of carbohydrates (sugar) from the tree. This was discovered more than 50 years ago. Saplings were subjected to light, and it was found that more light had a positive effect on the mycorrhiza (the saplings produced more sugar), while increased nitrogen fertilization had a negative effect. The sapling does not need the help of the fungus to absorb nitrogen if there is a nitrogen surplus. We know now that this is a very simplified description of the process.

What role, then, does acidification play? Different fungi have different requirements on the acid concentration in the soil. This is measured using pH units. A change in the pH value from 7 to 6 increases the degree of acidity tenfold, and from 7 to 5, a hundredfold, and so on. Neutral soil and clean water has a pH value of 7. Chanterelles grow best when the pH equals 5.5, a value that is very close to the pH of a normal birch grove. An increase in "acid rain" (sulphur fallout) has a negative effect on the soil's pH from the perspective of the chanterelle. A study in the Netherlands conducted from 1976 to 1985 measured the air's sulphur dioxide content (another measure of acidity) at a few hundred sites, and correlated this with the presence of chanterelle fruitbodies before and after 1975. The share of sites where chanterelles still remained after 1975 decreased in relation to the acidification level.

However, acidification is not the only factor in the Netherlands. It is known that there has been a drastic increase in the nitrogen level in the soil as a result of fertilization.

Forests once rich in herbs are now abundant in blueberry bushes, while *Cantharellus tubaeformis,* which prefers more acidic soil, may have become more common. Different fungi require different soil qualities. There are species that primarily grow on sites where the soil is high in lime. This applies, for example, to the Horn of Plenty (*Craterellus cornucopioides*).

In conclusion, we would like to briefly discuss the geography of the chanterelle. As chanterelles form ectomycorrhiza, it is interesting to note the distribution of ectomycorrhiza in the world. This is part of a map by

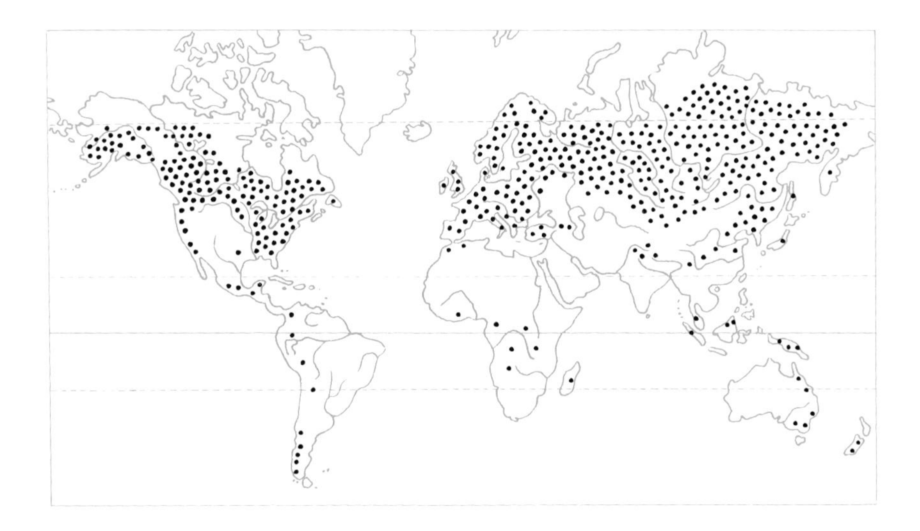

Areas where ectomycorrhiza dominates (marked); from Moser (1967)

Moser. It shows areas in the world where ectotrophic mycorrhiza is dominant. In these areas one can expect to find various kinds of chanterelles. Yellow chanterelle species can, for example, be found in large parts of the northern coniferous region. We will discuss this in more detail when we later discuss the various species.

Popular Names for Chanterelles

Chanterelle is derived from the Greek word *kantharos*, a kind of drinking vessel, via the word *cantharella* in modern Latin, from *cantharellus*, or "a small goblet." During the eighteenth century, the word *chanterelle* was introduced to Sweden through French. The modern Swedish name is *Kantarell* and the modern English name, of course, is chanterelle.

Popular names are closely tied to cultural history and to customs. The reason that there is no Swedish or Anglo-Saxon popular name for the chanterelle is probably due to the fact that Swedes and Anglo-Saxons rarely ate mushrooms in the past. In addition, the chanterelle was not linked to superstition, as was the case with some other mushrooms. The chanterelle was late to be included in cookbooks (see pages 88–89). The Danes have the name *vingesvamp* (wing mushroom) which refers to the often wavy edge of the cap; see Buchwald (1937). The Finns have the word *keltasieni* (yellow mushroom); see Edelmann (1989).

On the other hand, in continental Europe the chanterelle has many popular names, which indicates that it has been used for a long time. For example, in 1841 Krombholz wrote, *"Er wird allgemein genossen"* (it is commonly eaten). Even Clusius (see page 16), who was interested in popular names for mushrooms, noted the German name *Reheling* on a color plate in *Codex*. He also mentions the Hungarian name *Niwl gomba*, which means "hare mushroom."

Many dialectal chanterelle names in German, French, and Italian, as well as in some Slavic languages, refer to the yellow color, either directly or indirectly (as a yellow animal). Other animal names may also be included. Below are examples of chanterelle names in various languages:

German: *Dotterpilz* (egg yolk mushroom), *Eierschwamm* (egg mushroom), *Gelbhähnel* (yellow chick), *Gelbling* (yellowing), *Galuschel* (yellow ear), *Hasenöhrlein* (little hare ear), *Hühnling* (chick), *Pfifferling, Rehling, Rehfüsschen* (*Reh* means deer, *Fuss* means foot).

The German name *Pfifferling* refers to the somewhat sharp taste of the raw mushroom (*Pfeffer* means pepper).

A very detailed account of German chanterelle names can be found in Marzell (1943).

French: *crête de coq* (cock crest), *gallinace* (chicken), *girolle* (from the Old French word *girer*: to twist, referring to the shape), *jaunette* (*jaune*: yellow), and *oreille de lièvre* (hare ear). *Girolle* and *chanterelle* are the two most common names for chanterelle in French.

Italian: *capo gallo* (cock crest), *finferlo* (in south Tyrol *finferli*, related to the German *Pfifferling*), *gallinaccio* (chicken), *galletto* (young rooster), and *orecina* (little ear).

Some words mentioned from the Slavic languages are the Russian name *lisitjka* (fox mushroom) and the Czech names *liška* (also "fox mushroom") and *kuřatko* (chick).

The Dutch use the names *Hanekam* (cock crest) and *Dooierzwam* (*Dooier*: egg yolk).

Only a few of the other chanterelles and chanterelloid mushrooms have popular names. Long ago, *Gomphus clavatus* was referred to as *Schweinsohr* (swine ear) in German literature. *Craterellus cornucopioides* is referred to as *Totentrompete* (trumpet of death). That name is also found in French as *trompette de la mort* and in Italian, *trombetta da morto*. Of other chanterelloid mushrooms, *Hydnum* fungi early on received popular names. In German they are referred to as *Stoppelpilz* (stubble mushroom) or *Semmelstoppelpilz* (*Semmel* means French roll), and in French they are called *barbe de vache* (cow beard) or *pied de mouton* (sheep's foot).

Chanterelles and Chanterelloid Fungi

On the following picture pages and in the text, we will not only discuss *Chanterellus* and *Craterellus*, but also some other species that previously were considered chanterelles (i.e., chanterelloid, or chanterellelike) mushrooms. All of these species occur in Europe and most of them are commonly found in North America. We will also give some examples of chanterelles unique to Africa and North America.

In some mushroom floras (e.g., the French *Flore analytique des champignons supérieures* by Kühner and Romagnesi from 1953), both chanterelles and *Craterellus* fungi are referred to as *Cantharellus*, while they are classified in the subgenera *Cantharellus* and *Craterellus*, respectively. The distinguishing difference is whether they have clamp connections or not (see page 12). Fungi belonging to the group *Cantharellus* have clamp connections. The hymenium, the basidium-carrying tissue layer, can either be wrinkled or smooth within the same group. *Cantharellus aurora*, with only a slightly veined hymenium, has clamp connections and is now considered to belong to *Cantharellus*. *Craterellus cinereus*, however, has a wrinkled hymenium and no clamp connections and is now considered to belong to *Craterellus*. When the genus *Craterellus* was established, it was based on the appearance of the hymenium, if it was almost smooth. In the book, *Sveriges ätliga och giftiga svampar* (Edible and Poisonous Mushrooms of Sweden), *Craterellus clavatus* is named so because of its almost smooth hymenium. Modern data based on DNA studies implies a different system, where *Craterellus* include *Pseudocraterellus*, *Craterellus*, and former *Cantharellus* species with thin flesh, such as *C. tubaeformis*.

What is most important? Today, an effort is being made to find groups with genetic relationships, thus reflecting evolution over time. DNA analysis is the most powerful tool for such studies. Older systems aimed at gathering look-alikes based on morphology and later on microscopal characters such as cell division. The mechanism of cell division among chanterelles was not discovered until the beginning of the twentieth century by the Swede H. O. Juel. Both chanterelles and *Craterellus* fungi have stichobasidia, basidia where the cell division occurs in the lengthwise direction of the basidium (see page 12). This brings the genera close to each other. When we include *Hydnum* fungi in the same family as the chanterelles, we do so because they have stichobasidia.

Studies of the cell division of *Gomphus clavatus* have shown it to be closely related to the genus *Clavariadelphus* (see page 78). Both have clublike basidia with transversal cell division—chiastobasidia (see page 12). *Gomphus clavatus* is now included in the chanterelloid genus *Gomphus* but in a separate family.

When we join genera to a family, the family gets its name from one of the genera in the family, but the suffix *-aceae* is added. The chanterelle family may therefore be called Cantharellaceae after the genus *Cantharellus*. If the genus *Hydnum* is included, the family name Hydnaceae would also be acceptable. There are rules for nomenclature, but they change every now and then. Sometimes well-established names are kept, even if they aren't current. Decisions about this are usually made at botanical conferences. Some rules regarding nomenclature are discussed on pages 114–115.

For those who pick chanterelles for cooking, this discussion is, of course, of little interest. For them, the important issue is how they taste and how easy it is to find them. It is more important to know how they grow and how available they are from site to site. If one learns that the yellow chanterelle found in Europe has a rich aroma, and that the chanterelle found, for example, on the North American west coast is almost odorless, it brings up the question of whether this actually is the same mushroom. Modern techniques using DNA analysis have confirmed the hypothesis that *Cantharellus formosus* is the commonly occurring Pacific Golden Chanterelle, formerly believed to be the same as the European *Cantharellus cibarius*. Since people need to know what they are eating, and since odor, size, cultivation techniques, host trees, ecology, pigments, and canning qualities are different, correct scientific names are important to distinguish these species. But the question still remains: How much may a characteristic in a species vary, and how many characteristics may be different before we divide one species into two? The question, in many cases, is still open.

Cantharellus cibarius

When Linnaeus visited Västergötland, Sweden, he made the following notation on July 21, 1746: "Large numbers of the chanterelle or *Agaricus caulescens, lamellis decurrentibus,* an edible mushroom, grew in great quantities on Hunneberg, and it was different from other fungi of its genus, as its lamellae are branched, which is unusual among these fungi."

Linnaeus had adopted the German Johann Jakob Dillenius's classification of fungi into those without a cap and those with a cap (pileus). The capped mushrooms were divided into mushrooms with lamellae and those without. The lamella-equipped fungi were divided into edible *(esculentes)* and inedible *(noxii)* fungi.

Based on what we know today, Dillenius's classification seems primitive. The lamellae of the chanterelles are not really lamellae but richly furcated ridges or veins. Linnaeus had adopted the Latin genus name, *Agaricus*, from Dillenius. The name of the species, *caulescens* (Latin, meaning equipped with a foot), was later changed by Linnaeus to *cantharellus* (from the Greek word *kantharos,* meaning goblet). The current Latin name, *Cantharellus cibarius* (from the Latin word *cibus,* meaning food), was coined in 1763.

Alpine birch forest (Betula pubescens *ssp.* tortuosa) *in northern Sweden with an inlay of dwarf cornel* (Cornus suecica) *in fall colors*

The chanterelle has an aromatic scent that reminds us of apricots. It is therefore possible to teach dogs to find chanterelles in the same way they are taught to locate concealed molds, or truffles.

In Sweden, the chanterelle is rarely infested by insects. However, in Canada, chanterelles have been found infested with maggots.

Chanterelles often grow in clusters in coniferous forests rich in moss, but they can also be found in birch forests in mountainous areas among grasses and low-growing herbs. In central Europe, chanterelles often grow in beech forests, where other similar species and forms can also be found (see pages 38 and 42). Sometimes one can find totally white chanterelles. This can be the yellow chanterelle which has lost its pigmentation. *Cantharellus cibarius* is thought to be a cosmopolitan species. However, this scientific name has been, and may still be, used by mistake for several yellow chanterelles such as *Cantharellus formosus* (see page 84) of western North America. *Cantharellus cibarius* seems to occur in eastern North America but occasional findings are reported also from the west coast. The only chanterelle cultivated at experimental scale is the European *C. cibarius*.

On the next page we show the chanterelle's variations in shape. The pictures will speak for themselves. Notice, however, the pronounced ridges and folds on the underside. The mushrooms shown were picked in the Runby forest in Upplands Väsby (Sweden).

Chanterelle *Cantharellus cibarius* Fr.

Cantharellus pallens

Fifty years ago, in the birch and oak pastures in the archipelago outside Stockholm, Sweden, one could find large numbers of this fleshy, first cream-colored and then pale yellow chanterelle. It first pops up around midsummer, after mild early summer rains. It can still be found in parks and deciduous forests, but it seems to be limited to better soil with trampled-down vegetation. It becomes rarer in less open areas.

As early as 1916, Waldermar Bülow noted that "a pale yellow form can also be found, with a very thick cap and a stem which is about 2 centimeters thick."

This different-looking chanterelle must have been noticed earlier, but it was not until 1959 that it was given a specific name, *Cantharellus pallens* (the Latin word *pallens* means paling), by the Czech scientist Albert Pilát. He wrote (freely translated), "This form, found in deciduous forests, was not given a name of its own for a long time, as it was thought to be only a form influenced by the deciduous forest and the type of soil found there." Pilát became convinced that this mushroom was quite separate from the ordinary chanterelle when he picked it from a twenty-year-old spruce culture, where he also found ordinary chanterelles.

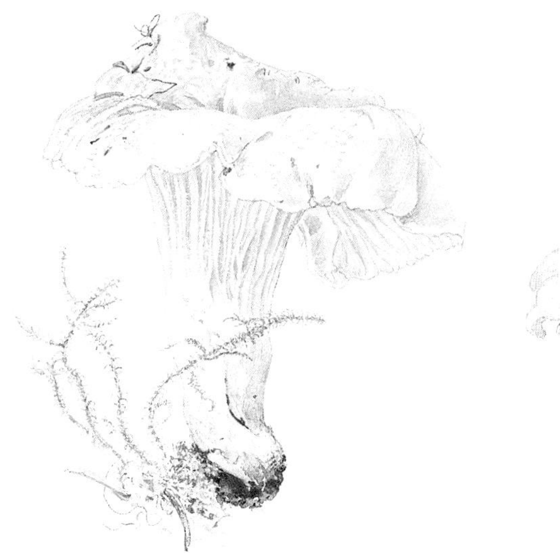

Chanterelle (white form) from Medelpad (Sweden)

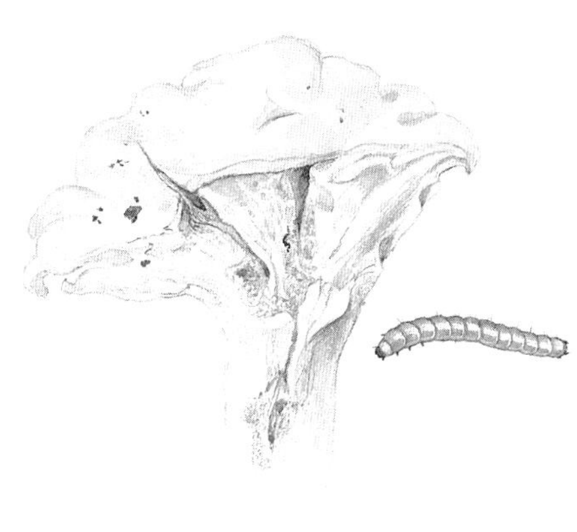

Pale chanterelle infested with an Elateridae larva (click beetle)

It is difficult to define the species, as nobody has attempted to cross *C. cibarius* with *C. pallens*. However, it must be noted that it is easy to recognize the pale chanterelle and that its ecology and distribution differ from that of the ordinary chanterelle. Also, DNA studies support the separation into two species. The all-white, probably unpigmented variety of the normal chanterelle, which can be found in coniferous forests north of the northern limit of oaks, is clearly different from the fleshy mushroom found in deciduous forests. It is probably this mushroom that the Czech Velenovsky referred to as early as 1920 as *Cantharellus pallidus*. He wrote, "Similar to *Cantharellus cibarius* . . . rare in coniferous forests." The illustration to the left above shows an almost-white chanterelle that was found in a coniferous forest in Sweden north of the oak tree line. It has not been determined whether this is a form of a yellow chanterelle or a separate species. The Elateridae larva is found occasionally in chanterelles.

The pale chanterelle has been found as far north as the oak tree line. In Sweden, it is primarily found in culturally impacted areas, most often in the vicinity of hazel. It has also been called hazel chanterelle. It is widely distributed but rather uncommon in Europe. A similar species, *Cantharellus subalbidus*, is found in North America (see p. 85).

The mushrooms depicted were picked in the Runby forests in Upplands Väsby, and the mushroom site in Algutsrum in Öland (Sweden).

Pale Chanterelle *Cantharellus pallens* Pilát

Cantharellus amethysteus

In markets and vegetable stands in western Europe, one finds, every now and then among the chanterelles, mushrooms with a violet-hued cap covered with small scales—*Cantharellus amethysteus*. It can be fleshy and robust or thin and slender. From a culinary standpoint it is not at all different from the ordinary chanterelle. The aroma is the same and the fat-soluble flavorings contribute to a delicious chanterelle butter.

The color variations among chanterelles are sometimes mentioned in older books, but the amethyst coloring is not described until 1882 by the Frenchman Quélet, who classified the mushroom as a variety, *Cantharellus cibarius* var. *amethysteus* (from the Greek word *amethysteos*, meaning amethyst, wine-colored). An English mushroom flora from 1922, by Carleton Rea, lists it as a separate species, *Cantharellus amethysteus*. He notes that it is not uncommon in England, but the Kew Herbarium provides evidence of it growing in the southern parts of the country.

Cantharellus amethysteus can be found in beech forests, but also near

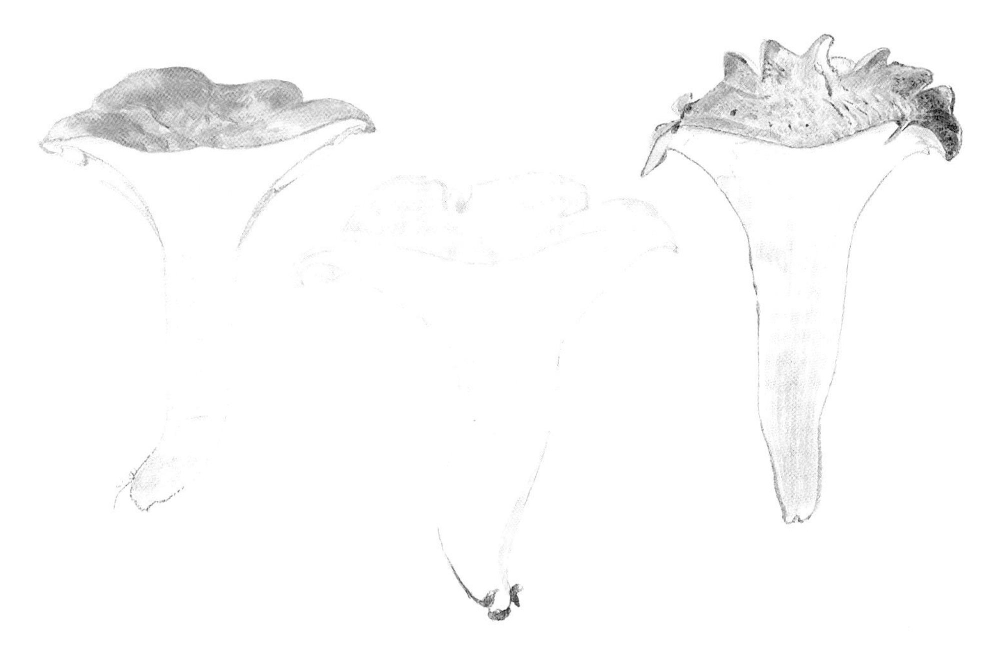

Plates depicting a yellow chanterelle, a pale chanterelle, and an amethyst chanterelle

aspen groves at the edges of meadows in the Alps. The clay soil is loamy. It often grows in barren mineral soil.

Cantharellus amethysteus has also been found in areas of central Germany (Harz, Coburg, etc.). It has been reported as rare in Italy, but is so common in Austria that it has colloquial names in German (e.g., *Blauling*, or "blue one"). In French books it is referred to as a mushroom found in mountainous areas (*montan*). A special study of the mushroom flora in Saarland (on the border of France) resulted in a 103-page map, of which seven pages contain notes of this kind of chanterelle.

Young amethyst chanterelles are completely violet in color. Not until the mushroom has developed does the surface of the pileus crack and become covered with small scales. A yellow chanterelle, a pale chanterelle, and an amethyst chanterelle are shown at the top of this page.

The landscape on the next page is at an altitude of between 600 and 700 meters above sea level, in Buch, close to the Inn River in Austria. It is an experience to capture a beautiful nature scene, but being an artist is not that easy—anything can happen. A watercolor artist is dependent on water; however, the end result of the artist's attempt to capture the site where the amethyst chanterelle grew ended on a good note. The sketch was done when a roaming dog licked up the last drops of the water supply.

Amethyst Chanterelle *Cantharellus amethysteus* Quélet

Cantharellus friesii

The Swedish poet Jeanna Oterdahl once wrote, "Have you seen Mister Chanterelle, over there on the juniper hill, he came here night before last with his cap pushed back. It is yellow and it is fine. . . ." This well-known poem reflects our view of the chanterelle here in Sweden. To find, in a deciduous forest, after traveling thousands of kilometers south to the Alps, a golden *Cantharellus friesii* a few centimeters high, shining among moss, *Polygonatum odoratum, Solidago virgaurea, Luzula luzuloides*, and *Melampyrum pratense* is a wondrous experience.

Cantharellus friesii was considered a separate species for the first time in 1869 by the mycologist Quélet, who coined the Latin name in honor of the Swedish mycologist Elias Fries. Just like *Cantharellus cibarius*, it has a faint smell of apricots and, when raw, a mildly bitter taste. It is very characteristic with its somewhat velvety, dark orange cap, which can reach a width of 3 to 4 centimeters, becoming paler toward the edges. The middle is initially lower. The German name is *Samtpfifferling* (velvet chanterelle). The stem is also orange, and the flesh has a pale pink coloring (see the

plate on page 48). This mushroom is rare north of the Alps, but common south of them, at elevations of between 500 and 1000 meters above sea level.

Its preferred conditions are not yet totally known. The specimens shown in the illustration were found close to the town of Jenbach in Osttirol, Austria. In Scandinavia, it has been found in three areas of the Norwegian Sörlandet. Findings of unusual looking chanterelles in Skåne, Sweden, may actually have been *Cantharellus friesii*. Like many other rare mushrooms (see pages 59, 63, and 71), *Cantharellus friesii* is sensitive and demanding. On the European mainland, this also applies to *Cantharellus cibarius*, which has been included on a list of endangered species in four countries. A small red chanterelle of the southern and eastern U.S. is *Cantharellus cinnabarinus*. Another fragile but colorful species is the bright yellow *Cantharellus minor*, which favors hardwoods in the U.S.

The picture on the next page shows another *Cantharellus friesii* site than the one where the material for the species page was picked. It is at Hall near Innsbruck, Austria.

Cross section of Cantharellus friesii

Orange Chanterelle *Cantharellus friesii* Quélet

Cantharellus tubaeformis

This thinly fleshed, trumpetlike chanterelle was probably not used as food before the beginning of the twentieth century. It is only during the last fifty years that it has become popular in the kitchen, and now it is even served in soups in many restaurants. Because of its long shelf life, it is one of the few wild mushrooms sold in Sweden today.

During the first half of this century mushrooms were in much greater supply in Europe. In the beginning of the century one could find nearly thirty species for sale in the market of Munich alone. In 1901, between 70 and 90 metric tons of ordinary chanterelles were sold, but no *Cantharellus tubaeformis* (Duggar, 1920).

Even in Sweden, large volumes of mushrooms were sold in the markets until the 1940s. At that time, both *Cantharellus tubaeformis* and *Craterellus cornucopioides* were already popular. However, about one hundred years earlier, Krombholz wrote, ". . . not much used, although both taste and scent are pleasant."

The dull color and the mushroom's thin flesh has discouraged pickers in the past. It is not included in mushroom books from the 1860s.

Yellow form of Cantharellus tubaeformis *and cross sections of* Cantharellus tubaeformis *and* Cantharellus aurora.

Cantharellus tubaeformis has confounded mycologists. In older literature it was perceived as two separate species, one with a totally hollow stalk and the other with a partly hollow stalk. This mushroom has also been confused with *Cantharellus aurora* (see page 55). In 1874, Fries listed two different species in *Hymenomycetes europaei*: *Cantharellus tubaeformis* (*tubaeformis* means trumpetlike) with a hollow stem, and *C. infundibuliformis* (*infundibuliformis* means funnellike). Today, they are considered a single species. The oldest name, *C. tubaeformis*, is the valid name.

Cantharellus tubaeformis belongs to a certain group of chanterelles, *Leptocantharellus* (from the Greek word *leptos*, meaning thin). Another name for the group is *Phaeocantharellus* (from the Greek word *phaios*, meaning dark or gloomy) based on its dark color. However, *Cantharellus tubaeformis* can be totally yellow, *C. tubaeformis* var. *lutescens* (from the Latin word *lutescens*, meaning yellowing), see the picture above. According to recent DNA studies, *Cantharellus tubaeformis* belongs to the genus *Craterellus*, together with other members of Cantharellaceae with thin flesh.

The species *Cantharellus tubaeformis* grows both in coniferous and beech forests in large areas in Europe. It is also common in North America, often under its old name, *Cantharellus infundibuliformis*. The color of the spores is usually white but may also show a yellow variation.

The mushrooms depicted on the next page were picked in the Runby forest in Upplands Väsby. This forest is also illustrated.

Funnel Chanterelle *Cantharellus tubaeformis* [Bull.: Fr.] Fr.

Cantharellus tubaeformis

Cantharellus aurora

Just as was the case with the fungus *Cantharellus tubaeformis*, *Cantharellus aurora* was also not considered a good cooking mushroom until the beginning of this century. In his edited version of M. A. Lindblad's mushroom book from 1901, Lars Romell, Sweden's foremost mycologist after Elias Fries, summarized the characteristics of *Cantharellus aurora* and *Craterellus cornucopioides* as follows: "Both of these are hollow and their flesh is very thin, for which reason they do not have much value." On the other hand, Edmund Michael stated in his book *Führer für Pilzfreunde* (Guide for Mushroom Lovers), from 1901, that the fungus *Cantharellus aurora* is edible and tasty, but that it cannot be found everywhere. But where it can be found, there are large numbers of it. Elias Fries mentioned it in 1821, under the name *Cantharellus lutescens*, but changes it later to *Craterellus lutescens* due to its smooth, only veined, shining red and yellow hymenium; compare *Craterellus cornucopioides* (see page 66). Due to the fact that the species *Cantharellus aurora* has clamp connections (see pages

12, 32), and since we follow the system of Kühner & Romagnesi (1953), this species has been placed in the genus *Cantharellus*. Recently DNA data imply that it should belong to *Craterellus*. Its name was recently changed from *Cantharellus lutescens* to an older name, *Cantharellus aurora* (the Latin word *aurora* means red light of dawn).

This name also goes back to a book by the German Batsch, from 1783. This species is often confused with the fungus *Cantharellus tubaeformis*, perhaps because a yellow form of it has been described.

The species *Cantharellus aurora* grows in damp areas in lime-rich coniferous forests. In 1874, Elias Fries wrote about the mushroom's habitat "on damp sites in coniferous forests in mountain areas." However, no mention is made of the fact that the fungus prefers soil rich in lime. Bruno Henning's revised version of Michael's above-referenced book, *Handbuch für Pilzfreunde* (Handbook for Mushroom Lovers), from 1964, contains a mention that the mushroom prefers "coniferous forests in the mountains, and mossy floors, especially rich in lime." Earlier Fries had mentioned the mushroom's strong fruity flavor, although he rarely commented on scents in his descriptions. According to hearsay, the reason for this was that he could only sense strong smells because he used snuff.

This mushroom, just like *Craterellus cornucopioides*, is very suitable for drying, and because of its aroma it is good in soups and sauces. In the lime-rich areas of central and southern Europe, it is common and even more popular than *Cantharellus tubaeformis*. In Scandinavia, it can mostly be found in areas where the soil is rich in lime, as far north as the polar circle. During some years, hundreds of liters of the mushroom have been picked locally as far north as Jämtland, Sweden. The species is widely distributed in North America.

Both *Cantharellus tubaeformis* and *Cantharellus aurora* sometimes lose their pigmentation and become totally white. Recently a new species within this group of fungi was found in the north of Norway. The color pigmentation of this fungus is similar to the pigmentation of *Cantharellus tubaeformis*, but the hymenium is smooth. This fungus was named *Cantharellus borealis* (the Latin word *borealis* means northern).

The pictures of the mushrooms and their environment are from Upplands Väsby, Sweden.

Yellow-stemmed Chanterelle *Cantharellus aurora* [Batsch] Kuyp.

Cantharellus melanoxeros

This mushroom, whose color changes from yellow to dirty yellow, with clear, yellow-grey to grey-purple furcated ridges on the underside of the cap, is very rare. The fungus turns dark when handled or dried, and has therefore been given the Latin name *Cantharellus melanoxeros* (the Greek word *melas* means black and *xeros* means dry). One species found in central Europe goes by the more recent name *Cantharellus ianthinoxanthus* (from the Greek *ianthinos*, meaning purple, and *xantos*, meaning yellow) and has been considered a different species from *Cantharellus melanoxeros*, but these fungi seem so close to each other that separating them into two species does not seem necessary.

The species *Cantharellus melanoxeros* is also included in the Frenchman Desmazières's herbarium from 1829 and was described in mycological literature the following year. Some illustrations from the latter part of the nineteenth century, which have other names, may also depict *Cantharellus melanoxeros*. One of these illustrations can be found in the book by the

Illustration from Bresadola's Fungi tridentini

Italian Giacomo Bresadola, titled *Fungi tridentini*, from 1881, shown above. Both the description and the picture point at *Cantharellus melanoxeros* because of the fleshy stalk and the color change. Bresadola referred to the fungus as *Cantharellus infundibuliformis* var. *subramosus* because he considered it a variety of *Cantharellus tubaeformis* (*infundibuliformis*). However, this mushroom is never as fleshy. No other related chanterelle is known to grow in Europe. Mushroomers have also sometimes mistaken the fungus *Cantharellus melanoxeros* for *Cantharellus cibarius*. It should certainly be edible, but as it is endangered, it should be left untouched. However, the word *endangered* is linked to the growing site rather than to the species. It is the site that should be protected.

Cantharellus melanoxeros grows in fertile soil in deciduous forests rich in clay and topsoil. Aside from a dozen Swedish sites, all of them south of the river Dalälven, there are seven other sites in Scandinavia, six of which are in Norway and one in Denmark. It has been found in many countries in western Europe (e.g., France, Italy, Switzerland, and Germany). Around twenty sites were marked in an atlas from 1991, by Krieglsteiner in Germany. It is not known in the U.S.

The pictures of mushrooms and their environment on the next page are from Männö in Södermanland (Sweden).

Blackening Chanterelle *Cantharellus melonoxeros* Desm.

Craterellus cinereus

This species was depicted in a book published in 1798 by the Dutchman C. H. Persoon, which included pictures and descriptions of lesser-known fungi, titled *Icones et descriptiones minus cognitorum* (see page 18). Krombholz depicted it in 1841, together with other chanterelles, in his work "Schwämme" (see page 20) and wrote that, due to its appearance and substance, it was not used. The Latin name used to be *Cantharellus cinereus* (*cinereus* means ash gray), but the mushroom does not have clamp connections (see pages 12, 32) and is therefore grouped with the genus *Craterellus.* It should therefore be called *Craterellus cinereus,* which is in accordance with DNA studies. It should also be mentioned in this context that mycological literature contains information about a species with clamp connections. The depicted specimens do not have clamp connections.

In 1879, Finland's foremost mycologist, P. A. Karsten, published the first comprehensive study of mushroom flora in Swedish. This was a translation of Fries's *Hymenomycetes europaei,* in which he wrote, "The cap is

Cross sections of Craterellus cinereus, Pseudocraterellus undulatus, *and* Craterellus cornucopioides

open or pierced, covered with scaly wool, soot-brown with a gray hue; the stalk is soot-brown with a black hue; the lamellae are ash-gray. Forest, rare. Scandinavia (Femsjö)."

Craterellus cinereus often grows in parks and groves, in soils rich with lime, but it has also been found in coniferous forests. The growing conditions required by the fungus still require more study. It is deceptively similar to *Craterellus cornucopioides* but with clear, differentiated lamellae. As with many other chanterelles, the scent resembles that of apricots.

The sites where the species *Craterellus cinereus* usually can be found keep changing more and more as the areas are used for roads and other types of construction. There are many countries in Europe where it is also rare, but in these countries it has been classified as an endangered species. It is seldom seen at mushroom exhibitions, but this may be because it is easy to overlook. It often grows together with *Craterellus cornucopioides*. In the U.S. it is widely distributed but uncommon. We now know it grows in a number of areas in Scandinavia. It can be found as far north as Stockholm and in southern parts of Finland. It has also been found in southern Norway.

The mushrooms and growing sites depicted on the next page are from a site in the southern part of Djurgården close to Stockholm. These mushrooms are usually not found for many years in a row.

Black Chanterelle *Craterellus cinereus* [Pers.: Fr.] Pers.

Craterellus cornucopioides

In 1821, Fries gave the Horn of Plenty the name *Cantharellus cornucopioides* (the Latin word *cornucopioides* is similar to a horn of plenty), but later referred to it as *Craterellus* (from the Greek word *krater,* meaning goblet), with the justification that the hymenium underneath the cap is smooth and shiny, or wrinkled.

In older literature, *Craterellus cornucopioides* is not referred to as a cooking mushroom. Fries did not include it in his book, *Sveriges ätliga och giftiga svampar* (Edible and Poisonous Mushrooms in Sweden), but Strömbom wrote in 1881 that it was a good mushroom for cooking, although rare. And it was praised in a Danish mushroom book by Fabritius Buchwald from 1937. Morten and Bodil Lange wrote in the book *Gode spisesvampe* (Good Edible Mushrooms [1979]) that when the dried mushroom is chopped and mixed in, it adds a wonderful taste to sauces or stews. It is highly appreciated in the Mediterranean region.

Craterellus cornucopioides can be found both in deciduous and coniferous forests in Europe and North America. In central Europe, it is pri-

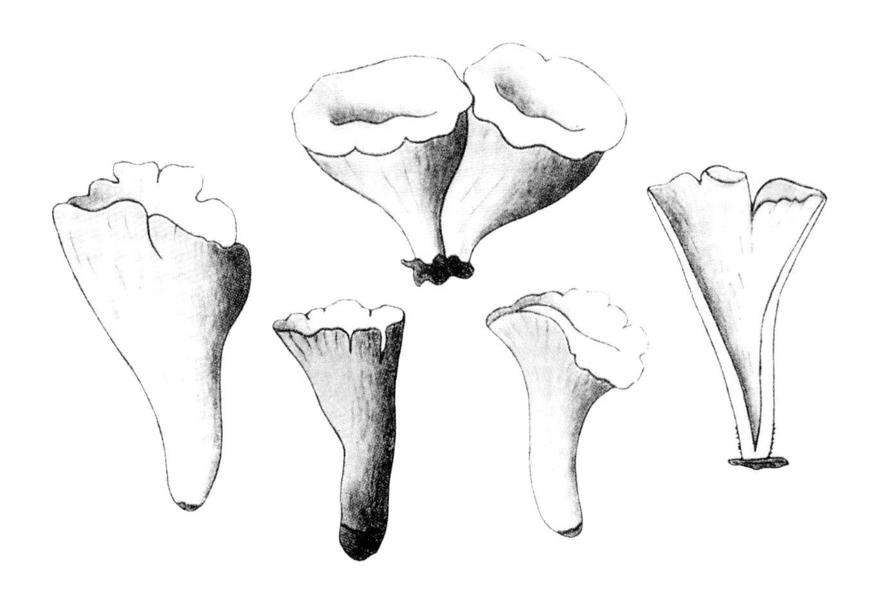

Picture of Craterellus konradii *from* Icones selectae fungorum

marily found in beech forests. In the rural areas of central Sweden, it is found in oak groves and on hazel-covered slopes. In the north it grows in dense spruce forests. Its northern limit of distribution is not known, but the mushroom can be found along the coast of Norrland, Sweden, and the west coast of central Norway. It is already rare in central Finland.

Craterellus konradii (named after the French mycologist P. Konrad) is a little yellow mushroom that resembles *Craterellus cornucopioides*. It is very rare and was described as growing in Switzerland and the French part of the Jura mountain range by the Frenchmen René Maire and H. Bourdot in 1927. It is depicted in the work *Icones selectae fungorum*, by Konrad and Maublanc, from 1924 to 1935. Some of the illustrations from that book are reproduced at the top of this page. Very few sites of this species have been reported but, in 1988, it was found in southern Norway and, in 1994, in Sweden. It is assumed that *C. konradii* is a variety of *C. cornucopioides* that has lost its black pigment.

Craterellus fallax is a North American species similar to *Craterellus cornucopioides*. Its hymenium is salmon or yellow tinted, and it is commonly found in southern and eastern U.S.

The mushrooms and the site depicted on the next page are from Upplands Väsby. *Craterellus cornucopioides* often grows in the same ecological environment as *Craterellus cinereus* (see page 63).

Horn of Plenty *Craterellus cornucopioides* [L.: Fr.] Pers.

Pseudocraterellus undulatus

Because of its diminutive size, this species is often overlooked. In addition, it grows in a similar ecological environment as that of *Craterellus cornucopioides* and *Craterellus cinereus*, and is often confused with them.

Similar species described in the beginning of the nineteenth century seem to be identical with *Pseudocraterellus undulatus*. Fries includes three different species in "Systema" from 1821: *Cantharellus sinuosus* (the Latin word *sinuosus* means rich in folds), *C. pusillus* (the Latin word *pusillus* means small size), and *C. undulatus* (the Latin word *undulatus* means wavy). Another name also was used in mycological literature: *C. crispus* (the Latin word *crispus* means rippled). Fries first listed this fungus as a variety of *C. sinuosus*, and later as a separate species. Mycologists now agree that all of these are variations of one species, which is now referred to as *Pseudocraterellus undulatus* (the Latin word *pseudo* means false). The name *Craterellus undulatus* can be traced back to a work from the eighteenth century by the Dutch mycologist Persoon. Recent DNA studies show that this species belongs to the genus *Craterellus*.

As indicated by the mushrooms in the illustration, the shapes of the species vary extensively. Specimens with an undulating cap border fit into the category *Cantharellus crispus*, while dwarf specimens fall under the category *C. pusillus*. They have as a common trait the smooth, veined, light-gray underside of the cap with its undulating edge. All the different variations have been found growing close to each other. They give off a faint odor. Mycological literature contains contradictory information regarding the nutritional requirements of the species. Some books state that the mushroom prefers soils rich in lime, while other books contend that it requires acidic soil. However, it seems to prefer rich soil, preferably meadows with deciduous hardwood trees. This mushroom can often be found in great numbers on each site.

Pseudocraterellus undulatus is widely distributed inthe U.S.A. but quite rare. It has been found in many western European countries, but the supply is clearly decreasing. Because of this, it should be protected and not used for cooking, although it is very tasty. It can be found in Sweden up to the river Dalälven in central Sweden, in the southwest corner of Finland, and on the Norwegian coast all the way up to Bergen. It has been found at about thirty sites in Denmark.

The mushrooms and the growing site on the next page were located in the Rydboholm castle park in Uppland, Sweden. However, the species does not always grow on flat ground. In Austria, we saw it in bare soil under tree roots, on extremely sloping, almost impassable terrain.

Wavy-capped Chanterelle *Pseudocraterellus undulatus* [Pers.: Fr.] Rauchert

Gomphus clavatus

The mushroom is called *violgubbe* ("violet-man") in Norwegian and *Schweinsohr* ("pig's ear") in German. Both names refer to color and shape: The violet underside and yellowish cap of the grown fungi resemble a pig's ear. Young mushrooms are club-shaped, and the species is considered to be related to the genus *Clavariadelphus* (*clava* means club; see page 78). As opposed to many other chanterelles, this fungus is fleshy and was, as early as 1821, listed by the Englishman S. F. Gray under its own genus, *Gomphus* (the Greek word *gomphos* means plug). Elias Fries first referred to it as *Cantharellus clavatus* (1821), but later he called it *Craterellus clavatus* (1836). In a book from 1798 by the Dutchman Persoon, it is called *Merulius violaceus*, a name that refers to the veined and purple underside. The name *Gomphus clavatus* is now quite accepted. In modern mycological literature it is listed under its own family, Gomphaceae, which is closely related to the clavarias. As opposed to the chanterelles, this genus has chiastobasidia (see page 12) and ochre spores. *Gomphus clavatus* is the only species of this genus in Europe, but there are several members of this genus in North America, including *Gomphus clavatus* and *Gomphus floccosus*.

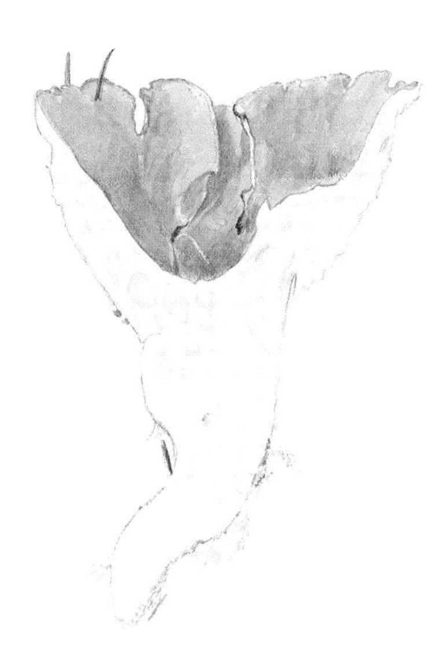

Cross section of Gomphus clavatus

Gomphus clavatus has been known as a popular cooking mushroom since long ago. In 1841, Krombholz (see page 20) noted that it was sold in markets in Prague, and Elias Fries included it in his work, (see pages 23, 25). He wrote, "Large numbers of this excellent and easily recognizable species can be found on damp sites in the forest. It prefers clay soil. It is not rare in central Sweden, but I have not been able to find it in Skåne or Småland, Sweden. The mushroom is prepared in the same manner as the chanterelle. In certain areas of Germany, such as Mähren, large volumes of it can be found in marketplaces. I find its flesh to be more spongy and not very tough."

In Scandinavia, the fungus has mostly been found in the northern part of Uppland, but it is fairly common in the area around Oslo, Norway. In addition, it has been found in the northern part of Finland, which indicates that its distribution is not yet entirely known. In western Europe it is not uncommon to find it in lime-rich mountain areas, but it is very rare in northern Germany, Denmark, and England. It appears to be distributed over large parts of the northern hemisphere including Pakistan, India, China, and Japan. It is fairly common in the coniferous forests on the North American west coast. There are many varieties of it, as well as of many other species within the genus *Gomphus*.

The mushrooms depicted were found among conifers in a forest in Storvreta, north of Uppsala, Sweden, and the site picture is from a coniferous forest in Uppland.

Pig's Ears *Gomphus clavatus* [Pers.: Fr.] S. F. Gray

Some Chanterelloid Mushrooms

False Chanterelle *Hygrophoropsis aurantiaca* (Wulf.: Fr.) Maire. Elias Fries stated in *Sweden's Edible and Poisonous Mushrooms*, regarding the ordinary chanterelle, "The only species with which it can be confused is the False Chanterelle, *Cantharellus aurantiacus*, which belongs among the hazardous mushrooms and should not be eaten." The word *aurantiacus* is Latin and means orange-colored. Later in the book, Fries wrote, "The species is commonly considered to be poisonous, although no clear proof has been presented in support of that assertion."

Today we know more about this species. It does not belong to the chanterelles, but is classified in a separate genus, *Hygrophoropsis* (the Latin word *hygrophoropsis* is similar to hygrophore), which is closer to agarics and boletes. The perception that the mushroom is not very suitable for cooking still stands. It is not currently recommended for cooking, but we do not know its chemical composition in detail. This mushroom does not establish mycorrhizal relationships, as chanterelles do, as it grows on decaying wood. Due to its close resemblance to the chanterelle, it is currently referred to as the False Chanterelle. It is commonly found in Europe and North America.

Truncated Club Coral *Clavariadelphus truncatus* (Quel.) Donk. This species is closely related to *Gomphus clavatus*. Its Latin name is derived from *clava*, or club, and *truncatus*, or truncated. It is also very similar to young specimens of *Gomphus clavatus*, but is considered part of the Clavaria family. *Clavariadelphus truncatus*, which grows in fir forests rich in nutrition, has a slightly sweet taste and is very suitable for cooking. It is widely distributed in Europe and North America.

Arrhenia auriscalpium (Fr.) Fr. This somewhat wrinkly, earlike mushroom, approximately one centimeter in size, often grows on mountains above the treeline. It can be found on recently thawed bare ground among thin mosses, and has therefore often been referred to as Moss Chanterelle. The Latin word *auriscalpium* means ladle with ear. Older mycological literature listed it among the chanterelles, but it has no relation to them.

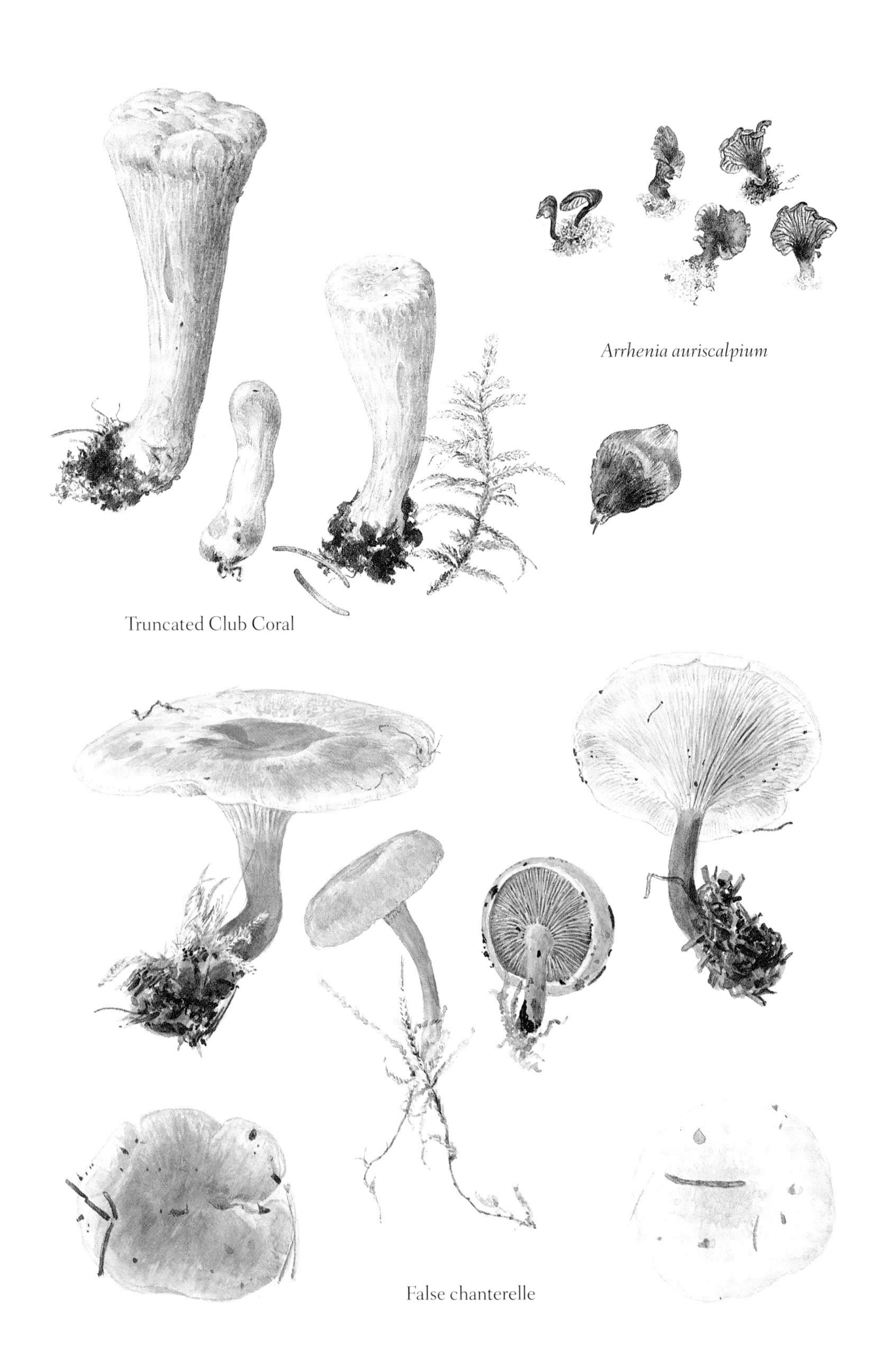

Arrhenia auriscalpium

Truncated Club Coral

False chanterelle

Hydnum Fungi (Hedgehog mushrooms)

In the introduction to chanterelloid mushrooms (see pages 32–33), we explained why the *Hydnum* species, or Hedgehog mushrooms, are grouped with chanterelles. Many mushroom pickers have sensed that they are close to chanterelles—in color, smell, and taste the resemblance is striking.

For a long time, these mushrooms have been popular in cooking, and they have been given many colloquial names (see page 31). These are fungi that cannot be confused with other mushrooms. Linnaeus already mentioned the species *Hydnum repandum* in his book *Flora suecica* (1755) and wrote, *"Erinaceus esculentus pallide luteus"* (hedgehoglike, edible, and pale yellow), and "grows in vast forests and is fairly rare." Even Fries wrote, more than a hundred years later (1861), that *Hydnum repandum* can easily be recognized by its pale yellow coloring, its fleshy, usually uneven and undulating cap, the surface of which is smooth and shiny. He also mentioned that the spines are usually of different lengths. Fries also wrote, "A subspecies *(Hydnum rufescens)* has a more slender stalk, a thinner, more irregular cap, with a fine woolly surface, and its spines are always of uniform shape. The color is reddish, flame-yellow." He continued by saying that there is no sharp distinction between this and the more common species.

Hydnum repandum has a wide distribution. Aside from beech forests, it can also be found in coniferous forests or birch forests. There it often grows together with the chanterelle. *Hydnum repandum* is no longer as rare as Linnaeus reported it to be in his day. Older German mycological books state that it is common in both deciduous and coniferous forests. However, over the last twenty years, it seems to have declined in Europe.

Hydnum repandum grows throughout Europe, North America, Japan, Tasmania, and Australia. Aside from *Hydnum rufescens* and *Hydnum repandum*, a very rare American species, *Hydnum albidum* (the Latin word *albidum* means whitish), has also been found in Europe. It was found on Gotland, Sweden, as recently as 1990. It was earlier found in Switzerland (1977) and on Mallorca, off the coast of Spain (1980).

Orange hedgehog *Hydnum rufescens* Schæff: Fr. Smaller and darker than *Hydnum repandum.* The cap has a reddish-yellow hue and is covered with a light fuzz. Its spines are pale orange and more irregular. It does not have the slightly bitter taste of *Hydnum repandum* but is still not considered as good for cooking. The Latin name is derived from *hydnon,* a Greek word for truffle, and the Latin word *rufescens*, which means blushing.

These mushrooms were collected from Salem and Upplands Väsby, Sweden.

Hedgehog *Hydnum repandum* L.: Fr. Badham writes in *Esculent Funguses of England* (1847) that *Hydnum repandum* is sold in Italy together with chanterelles. When cooked for a long time, it is delicious, with a slight oyster taste. The Latin species name *repandum* means turned upwards. See pages 88–89 for more about preparing chanterelles. In the U.S., the Hedgehog mushroom (sometimes referred to as *Dentinum repandum*) is becoming a popular choice.

These mushrooms were collected in Upplands Väsby, Sweden.

Hydnum albidum Peck. Mentioned in the introduction to the genus *Hydnum*. In Europe it has been found at four sites on Gotland, in eastern Jutland in Denmark, and at a dozen sites in western and southern Europe. It prefers soil rich in lime, and it grows in mossy coniferous forests near pine and spruce. The species epithet *albidum* means whitish. The cap is first almost as white as chalk, then cream-colored, and darkens if subjected to pressure. The stalk is often oblique and the upper side is a bit undulating. It occurs in the southern U.S.

The mushrooms shown below are from Kohlen's Mill in Visby, Gotland (Sweden).

Chanterelles Outside Europe

It is not easy to form a picture of the distribution of yellow chanterelles around the northern hemisphere. A map of the geographical distribution of the Swedish yellow chanterelle, *Cantharellus cibarius*, alone would require extensive study.

In the Pacific Northwest of North America, large numbers of yellow chanterelles grow in the temperate rain forests among Douglas fir on the Olympic Peninsula. It is a real adventure to climb in deep moss and across fallen tree trunks more than a yard in diameter. Suddenly one sees enormous 4- to 6-inch-wide chanterelles. But are they the same species as the European chanterelle? The coloring is more apricotlike and the surface of the cap is more velvety (see below) than the cap of the European *C. cibarius*. In addition, this mushroom does not have much odor. DNA studies

Pacific Golden Chanterelle Cantharellus formosus corner

White Chanterelle *Cantharellus subalbidus* Smith & Morse

show that the common Pacific Golden Chanterelle found in the Pacific Northwest of the U.S. is a species very different from the European *Cantharellus cibarius*. The correct name is *Cantharellus formosus* (see page 84).

However, large quantities of it are picked and sold at Pike Place Market near the port of Seattle, Washington, together with a giant species, *Cantharellus subalbidus* Smith & Morse (the Latin word *subalbidus* means somewhat whitish), which has a wonderful, intense apricot scent. This chanterelle is in high demand among local mushroom pickers, but European importers do not want it due to its white color. This is sad since its culinary value is higher than that for *C. formosus*.

The white chanterelle on the previous page is a separate species and has nothing to do with the European *C. pallens* (see page 38). *C. subalbidus* grows under Douglas fir, preferably in an undergrowth of salal (*Gaultheria shallon*), and American blueberry (huckleberry). Farther south—in California—there are also white chanterelles belonging to an undescribed species. They grow among "evergreen oaks."

On the eastern side of North America there is also a small species, *Cantharellus cinnabarinus* Schw. (the Latin word *cinnabarinus* means cinnabar red; see below). It is a species separate from *Cantharellus friesii*, which grows in western Europe (see page 46). It is enjoyed in cooking, but it is not as attractive as the white and the yellow chanterelles.

There is a commercial market for chanterelles not only in Europe and America. In several African countries many different kinds of chanterelles

Red Chanterelle
Cantharellus cinnabarinus Schw.

Kabengera *Kifundo*

are sold along roads and in markets. A few of these species are depicted above. The reddish chanterelle with a yellow underside, *Cantharellus longisporus* Heinem (the Latin word *longisporus* means long spores), is referred to by the local population in western Tanzania as **Kabengera**. It grows in dry forests among leguminous trees. We were told that this is the most common species. *Cantharellus congolensis* Beeli (the Latin word *congolensis* means from Congo) is an almost black species, but that is more rare. It is referred to as **Kifundo** by the local population. Both of these species are sold together with *Cantharellus cibarius*–like chanterelles.

Chanterelles can also be found in the southern hemisphere. In the highlands of Madagascar, many small yellow chanterelles are sold. It has not been determined if this species is identical to our own yellow chanterelle.

The illustrated specimens of *C. formosus* and *C. subalbidus* were picked among Douglas fir near Lake Crescent on the Olympic Peninsula outside Seattle. The illustrations of *C. cinnabarinus* and the African chanterelles were depicted according to pictures found in mycological literature.

Cooking with Chanterelles

Long ago, the ancient Romans appreciated mushrooms as a cooking ingredient. According to Fries (1836), Martial (43–104 AD) wrote, "Aurum et argentum facile est, Lenamque tonamque mittere; Boletos mittere difficile est" (it is possible to live without gold and silver, and one can resist the temptations of seductive women, but to abstain from eating mushrooms is difficult). Caesar's Mushroom, also known as Boletus (*Amanita caesarea*), was a favorite among the ancient Romans. One of the Emperor's gastronomes referred to it as *cibus deorum* (food of the gods). It was only served to the Emperor, and only on the most precious dishes (*boletaria*). The rest of the guests had to content themselves with *fungi suilli* (swine mushrooms). It seemed, however, that European interest in mushrooms was limited to the Slavic peoples and the Romans. The only Germanic people interested in mushrooms seemed to be those that inhabited the outer edges of the Roman Empire.

The chanterelle is not mentioned by name until the sixteenth century. Clusius (see page 16) mentions it as one of the edible mushrooms, "decimum-quatrum esculentorum fungorum genus" (the fourteenth edible mushroom). It is not referred to again as a separate species of edible fungi until much later. It was primarily the rural population that ate it, as indicated by all its colloquial names (see page 30–31).

In Sweden it was not introduced as an edible mushroom until the eighteenth century, through French cooking, and still not mentioned in cookbooks until much later. Linnaeus wrote in *Flora oeconomica* (1749), "1049 *Agaricus*. Suitable for cooking." (1049 refers to the chanterelle's

number in *Flora suecica*, 1745). *Swenska Economiska dictionnairen* (The Swedish Economic Dictionary) by Johan Fischerström (1779) contains the following passage: "*Agaricus Cantharellus* (Chanterelle) is included among the edible mushrooms, and is served at the tables of the nobility." Anders Jahan Retzius (1806) writes in *Flora oeconomica Sueciae*, "Fried in butter and eaten with salt and pepper, it is considered a delicacy by many," and Elias Fries (1836), "It is considered everywhere as one of the most important and best edible mushrooms, and it is also one of the few species which has been used in cooking in our country." In less than one hundred years, the attitude toward the chanterelle turned around. But mushrooms in general were still only prepared in castle kitchens and similar environments, as opposed to southern Europe, where they were enjoyed by the rural population.

Detailed instructions on how to prepare mushrooms in general, and chanterelles in particular, can be found in foreign cookbooks from the first half of the nineteenth century. In "Schwämme," Krombholz describes the culinary value of various groups of mushrooms. He notes that the chanterelle needs to be prepared well in order not to be hard to digest, and in *A Treatise of Esculent Funguses of England*, Badham (1847) writes, "The best way to prepare the chanterelle is probably to finely slice it in stews by itself or with other mushrooms." Margareta Nylander (1886), in her *Kokbok för husmödrar* (Cookbook for Housewives), recommends that it should be "cooked in butter."

It became apparent early on that all of the chanterelle's flavorings are soluble in fat. But it also contains water- and alcohol-soluble flavorings. Krombholz's pointers for preparing mushrooms have been used in some of the recipes on the following pages:

1. Cooked in butter and seasoned with salt, white pepper, and parsley.
2. Cooked in wine with coriander, Roman style.

3. Cooked in oil, Italian style.
4. Cooked with cream.
5. Cooked with bouillon.

It is important in all cases to release the flavor from the mushroom. Many chanterelle dishes can begin with the preparation of Chanterelle Butter, as follows:

Chanterelle Butter

Serves 4

½ to 1 pound (½ to 1 liter) fresh chanterelles
1 cup (250 g) butter

Preparation

If the chanterelles need rinsing, do so before cutting them, and let them dry on dish drainers or paper towels. Cut them into thin slices.

Cooking

1. Melt the butter on low heat in an enamel or stainless steel pan. Do not allow butter to brown.
2. Add mushrooms. Cover and let simmer until mushrooms contain no liquid and are golden yellow. Allow butter to cool slightly, and pour into a glass or porcelain bowl for further cooling.

Tips

The sautéed chanterelles are now ready for further use, and the cool butter tastes wonderful on newly toasted bread, crisp bread, or mild unsweetened crackers. It can also be used as a condiment with fish or mild meat (e.g., turkey). Butter-cooked chanterelles can also be lightly creamed in a frying pan.

Creamed Chanterelles

Serves 2 to 4

½ to 1 pound (½ to 1 liter) chanterelles, cooked in butter as above

¾ cup (2 dl) whipping cream or 1 cup (¼ liter) crème fraîche

Salt and white pepper

Cooking

1. Put the warm chanterelles in a frying pan, and allow to brown slightly over medium heat. The fat absorbed by the mushrooms during butter preparation should be enough so that none needs to be added during this step.
2. Add the cream or crème fraîche and a little salt and white pepper, and let it cook until the desired consistency is reached.

Tip

The chanterelles can be spread on fresh bread or used as a sauce.

Warm Mushroom Salad

Serves 4

½ pound (½ liter) mushrooms, such as chanterelles, shiitake, chestnut, and oyster mushrooms

3 ounces (90 ml) walnut or hazelnut oil

1 shallot, peeled and finely diced

1 clove garlic, finely chopped

Salt and freshly ground black pepper

3 young lettuces (choose 1 each from: lollo rosso, radicchio, curly endive, butterhead, cos, little gem)

Sprig of fresh tarragon, fennel or chervil, finely chopped

2 tablespoons (30 ml) raspberry vinegar

Preparation

Clean the mushrooms with a damp cloth and cut into manageable pieces.

Wash and dry the lettuces and place with the herbs in a large salad bowl.

Cooking

1. Heat a spoonful of oil in a medium-sized saucepan over moderate heat. Add the shallot and garlic and cook for about 3 minutes, until soft, but not colored. Add the mushrooms and cook until tender, about 3 to 5 minutes. Season with salt and pepper.
2. Pour the mushrooms with the garlic and shallot and pan juices onto the lettuce leaves. Heat the remainder of the oil and dribble over the salad. Finally, bubble up the vinegar in the same saucepan, reduce by

half, pour over the salad, correct the seasoning and toss well. Serve warm with French bread.

Soufflé of Chanterelles

Serves 6.

3 tablespoons butter
½ pound (½ liter) fresh chanterelles, sliced
3 shallots, minced
3 tablespoons all-purpose flour
½ cup (125 grams) half-and-half
¼ cup (62 grams) dry white wine
2 teaspoons chopped fresh tarragon
¾ teaspoon coarse salt
½ teaspoon white pepper
5 eggs, separated

Preparation

Preheat oven to 375 (180C) degrees.

Cooking

1. In saucepan, melt butter over moderate heat. Add chanterelles and shallots and cook, stirring frequently, until moisture is evaporated, about 15 minutes. Stir in flour, then gradually blend in half-and-half, and wine. Stir in tarragon, salt, and pepper, and cook over moderate

heat until thick, about 3 minutes. Stir about ½ cup (124 grams) hot mixture into egg yolks, then return to pan and stir to blend. Remove from heat.

2. Beat egg whites until stiff. Gently stir half of whites into mushroom mixture. Carefully fold in remaining whites. Pour mixture into buttered 2-quart soufflé dish and bake 35 to 40 minutes. Serve immediately.

Mushroom Sauce for Pasta

Serves 4

¼ to ½ ounce (10 g) dried Cantharellus tubaeformis
3½ ounces (100 g) fresh oyster mushrooms (Pleurotus *sp.*)
1 small yellow onion
1 strip lightly smoked bacon
1 cup (3 dl) soaking water
½ vegetable bouillon cube
1 teaspoon freshly crushed coriander seeds or freshly chopped cilantro
2 tablespoons butter
1 tablespoon flour
⅓ cup (1 dl) whipping cream

Preparation

Soak the chanterelles approximately 15 minutes.

Cut the oyster mushrooms into small pieces.

Finely chop the onion.

Cut the bacon strip into smaller pieces.

Pour the soaking water into a stainless steel pan to heat. Dissolve the bouillon cube in it.

Slice the soaked chanterelles.

Crush the coriander seeds.

Cooking

1. In a hot frying pan, sauté the onion in the butter without allowing the butter to brown.
2. Add the mushrooms and lower the heat. Sauté an additional 4 to 5 minutes.
3. Stir in the bacon.
4. Sprinkle with flour and dilute with water-bouillon mixture. Let simmer for 5 to 10 minutes.
5. Add the cream and let cook for a few more minutes.
6. Season with coriander or cilantro leaves.

Tips

The pasta should be cooked according to the instructions on the package. You can use fettucini, tagliatelle, or similar pastas. This kind of mushroom sauce can be found described in many mushroom books, but here we have given it a Roman touch by using oyster mushrooms, which have been used since Roman times. Coriander was also used in Roman kitchens.

Tagliatelle With Chanterelle Sauce

12 ounces (400 grams) chanterelle mushrooms

1½ ounces (50 grams) butter

1 tablespoon (15 mL) olive oil

1 shallot or green onion, sliced

Salt

Black pepper

1 tablespoon chopped parsley

1 lb (500 grams) green tagliatelle pasta

Preparation

Brush the soil off the mushrooms and wipe clean with a damp cloth. Do not wash them by immersing in water or you will ruin their taste and texture. Slice the mushrooms.

Cooking

1. Heat the butter and oil and gently fry the shallot or green onion until soft. Add the mushrooms and cook on a low heat for a few minutes. Season to taste and add the parsley. Keep warm while you are cooking the pasta in boiling salted water.

2. Drain the pasta and stir in the sauce. Serve at once.

Tip

Do not substitute other mushrooms in this dish as they are not as flavorsome as chanterelles.

Fettuccine with Sundried Tomatoes and Chanterelles

Serves 6

Sundried tomatoes, all the rage these days, were first developed by the Italians as a way to preserve vine-ripened tomatoes for sauce-making during the harsh winter and early spring months.

5 ounces (200 grams) unsalted butter

1 small yellow onion, finely chopped

Sea salt, to taste

6 ounces (225 grams) pumate, or 4 ounces (125 grams) dehydrated tomatoes

½ pound (1/2 liter) chanterelle mushrooms, cleaned and coarsely chopped

1 cup (250 grams) or a bit more heavy cream

1½ pounds (750 grams) fresh fettuccine, or 1 pound (500 grams) dried

Freshly ground black pepper

Cooking

1. Put the butter in a large skillet over medium heat. When the butter melts, add the onion and salt. Sauté gently until the onion is translucent.

2. Add the tomatoes, chopped into small pieces. Olive oil-packed tomatoes can be used without straining or pressing to remove the oil; the oil that clings just adds flavor. If you use dried ones, be sure not to disintegrate them by soaking or boiling them too long.

3. Add the mushrooms and sauté until they are tender, about 5 minutes.
4. Add the cream and reduce for about 3 or 4 minutes.
5. Cook the pasta al dente and drain, leaving it slightly wet.
6. If you have made the sauce ahead, reheat it now. Add the cooked pasta and toss well.
7. Serve on heated plates and top with plenty of pepper.

Award-Winning Mushroom Soup

Serves 4

½ ounce (15 g) dried Cantharellus tubaeformis

1 slice celeriac

1 piece of leek (2½ to 3 inches)

2 tablespoons olive oil

4 cups (1 liter) of water

1½ to 2 chicken or vegetable bouillon cubes

4 tablespoons flour

⅓ cup (1 dl) whipping cream

1 egg yolk

Parsley, minced, for garnish

Sherry, to taste

Preparation

Soak chanterelles in enough water to cover (15 minutes).

Drain the water and cut them into pieces (reserve the soaking water).

Cube the celeriac and slice the leek.

Cooking

1. Sauté the mushrooms in some of the oil.
2. Cook the mushrooms in the water (including the soaking water) together with the celeriac and the bouillon cubes in a stainless steel pan for 10 to 15 minutes.
3. Sauté the leek in the rest of the oil.
4. Pour the sautéed leek into a pan with a thick bottom and mix in the flour while stirring. Dilute with a little mushroom bouillon until the mixture is smooth without any lumps. Mix in the rest of the bouillon with the mushrooms, and let it cook for an additional 5 to 7 minutes.
5. Pour the soup into a bowl.
6. Stir in the cream mixed with the egg yolk when the soup has cooled down so that the egg yolk does not coagulate.
7. Add a little sherry and the minced parsley.

Tips

The soup can be seasoned in many different ways. Thyme and crushed juniper berries (6 to 8) give it a little game taste. Basil and sherry is another option, and red peppercorns and bay leaf is a third. The soup was created by Louis Bienen and received an award from the Swedish Society for Mushroom Promotion. Those who have access to *Craterellus cornucopioides* Horn of Plenty, can give a nice southern touch by using them in place of chanterelles, and then season the soup with freshly crushed coriander seeds or leaves.

Hubbe's Wilderness Stew

Serves 4

10½ ounces (300 g) of either hare, rabbit, or moose meat
and
10½ ounces (300 g) venison, thinly sliced
1 medium yellow onion, chopped
10 crushed juniper berries
1 pinch basil
1 pinch rosemary
Salt
White pepper
7 ounces (200 g) fresh chanterelles
1¼ cups (2½ dl) whipping cream
Parsley, minced, for garnish

Ingredients for marinade

¾ cup (2 dl) oil
2 yellow onions, chopped
1 bunch parsley, minced
White pepper
1 cup (3 dl) red wine

Preparation

Mix all ingredients for marinade.

Cut meat into pieces and marinate for 48 hours.

Cooking

1. Remove meat from marinade, place in a pan over medium heat, and brown until meat becomes golden brown. Transfer to a pot over medium heat.
2. Add onion and spices.
3. Season with salt and pepper.
4. Sauté the chanterelles in butter, and pour into the pan together with the cream. Let simmer until the contents of the pan thicken.
5. Sprinkle with minced parsley.

Tip

Serve this stew with pasta, or boiled or mashed potatoes.

The recipe was created by the restaurateur Hubert Förster, Dorotea, Sweden.

Hjördis's Pizza

Serves 4

Topping ingredients

9 ounces (240 g) venison, thinly sliced (1 package frozen)

10½ ounces (300 g) fresh or frozen chanterelles, cut into pieces

1 can crushed tomatoes

1 pinch thyme

Pizza crust ingredients

1 package (25 g) dry yeast

⅔ cup (1½ dl) warm water

½ teaspoon salt

1½ tablespoons olive oil

2 cups (4½ dl) flour

½ pinch black pepper

2 medium onions

2 tablespoons butter

¾ cup (2 dl) whipping cream

Salt

Oregano

Grated mild cheese

Preparation

Thaw the venison and the mushrooms, if frozen.

Pour the crushed tomatoes into a bowl, season with thyme and pepper, stir, and let sit.

Slice the onions.

Prepare the pizza dough when the reindeer meat has thawed.

Crumble the yeast into a bowl.

Pour water over the yeast.

Mix in salt and oil.

Stir in the flour and knead the dough.

Let it rise for 35 minutes.

Cooking

1. Sauté the onion in part of the butter while the dough rises, and put it on a plate.
2. Sauté the venison, separate it, cut it into small pieces, and add the sautéed onion.
3. Add cream and let it simmer until it becomes thick.
4. Add salt to taste and pour into a bowl.

5. Sauté the chanterelles in the remaining butter.
6. Roll out the dough to the size of a roasting pan or large pizza pan.
7. Sprinkle the crushed tomatoes, the venison, and the chanterelles on top of the dough.
8. Season with oregano.
9. Cover with grated cheese.
10. Bake at 450 (220C) degrees until the pizza is lightly browned.

Tips

The recipe was created by Hjördis Lundmark from Sundsvall's Mycological Association. She has also tried cut, freshly frozen chanterelles. As opposed to many other mushrooms, both the yellow chanterelle and the cultivated white mushroom seem to maintain their consistency quite well when frozen without first needing to be precooked. The cells in many mushrooms rupture if they are frozen without being precooked. Other mushrooms, such as Funnel Chanterelles and Hedgehogs, can also be used for pizza topping.

Marinated Hedgehog Mushrooms

1½ pounds (1½ liters) hedgehog mushrooms
(Hydnum repandum) *cleaned and cut into pieces*

Ingredients for marinade

1 to 1½ cups (3 dl) vinegar

⅓ to ½ cup (1 dl) oil

1 clove garlic

1 bay leaf

1 pinch thyme

1 teaspoon dried or fresh minced tarragon

1 tablespoon minced parsley

Preparation

Parboil the mushroom in a few quarts of water at low heat for 10 minutes. Pour out the water, and let the mushrooms drain and cool in a colander. Combine the ingredients for the marinade.

Cooking

Put the mushrooms in a bowl, and pour the marinade over them. The mushrooms must be allowed to cool for at least 5 hours before they are served.

Tips

Other mushrooms such as chanterelles can be marinated in the same manner. Marinated mushrooms can also be used as an appetizer. The recipe was created by Marguerite Walfridson in *Cooking Mushrooms and Mushroom Dishes* (Matsvampar och Svampmat, 1992).

Stuffed Chicken Breasts with Chanterelles

Serves 4

4 boneless chicken breast halves

4 thin slices prosciutto

Salt and black pepper

2 tablespoon olive oil

1 tablespoon butter

1 garlic clove, minced

½ pound chanterelles, cleaned and sliced

Juice of ½ lemon

¼ cup chopped fresh Italian parsley

Preparation

Preheat oven to 250 (130C) degrees .

Pound chicken breasts to flatten. Insert finger between skin and meat to loosen and insert slice of prosciutto in each. Season generously with salt and pepper.

Cooking

1. Heat oil and butter in a large skillet, preferably nonstick, over high heat.
2. Sauté the chicken until lightly browned on both sides, about 3 minutes per side.
3. Transfer to an ovenproof platter and keep warm in oven.
4. Reduce heat in pan to medium-low and quickly fry garlic, being careful not to burn.
5. Add mushrooms and sauté over medium heat, shaking pan frequently, about 5 minutes.
6. Sprinkle lemon juice and parsley and remove from heat. Scatter mushroom mixture over chicken and serve.

Veal Scaloppine With Chanterelles

Serves 4

1 pound (500 grams) thinly sliced veal scallops

Flour

Salt and freshly ground pepper

5 tablespoons butter

1 tablespoon olive oil

¼ cup (60 grams) shallots or onion, minced

1 clove garlic, minced

½ pound (250 grams) chanterelles, stems trimmed and chopped

1½ cup (125 grams) dry white wine

3 tablespoons chopped fresh Italian parsley

Preparation

Lightly coat veal with flour and pat off excess. Season all over with salt and pepper.

Cooking

1. Melt 1 tablespoon of butter and oil together in large skillet over medium-high heat.
2. Sauté scallops in batches, less than 1 minute per side, and transfer to platter.
3. Add 1 tablespoon butter to pan and reduce heat to medium.
4. Cook onion or shallots and garlic until soft.

5. Add mushrooms, turn up heat, and cook another 5 minutes, stirring frequently.
6. Pour in wine and cook until liquid is reduced to about 2 tablespoons. Reduce heat to low.
7. Cut remaining butter into pieces and swirl into pan one at a time until melted.
8. Stir in parsley and return scallops to pan. Cook a minute or so longer, turning veal to heat through, and serve.

Grilled Striped Bass With Chanterelles & Haricots Verts

Serves 6

The mushrooms

1 1/3 cups (320 grams) olive oil

2 pounds (2 liters) fresh chanterelles, cleaned

1/4 teaspoon salt

1/4 teaspoon freshly ground pepper

1 1/2 tablespoons chopped fresh thyme

4 large cloves garlic, finely chopped

4 large shallots, finely chopped

3 tablespoons sherry vinegar

4 tablespoons fresh lemon juice

The striped bass

6 (7-ounce/200-gram) portions striped bass, scaled and filleted

Olive oil

Salt

Freshly ground pepper

The haricots verts

½ cup (175 grams) butter

3 shallots, diced

1½ pounds (1½ liters) haricots verts or pousse pieds, blanched

Salt

Freshly ground pepper

1 tablespoon fresh lemon juice

Garnish

1 bunch chervil

6 lemon wedges

Preparation

Heat ⅔ cup (160 g) of olive oil in a sauté pan until it is hot. Add the mushrooms, salt, and pepper. Cook for 5 minutes over medium-high heat. Add the thyme, garlic, half of the chopped shallots, and the vinegar, and sauté for 1 more minute. Remove from the heat and let the mushrooms cool. Add the remaining ⅔ (160 g) cup of olive oil, the remaining chopped shallots, and the lemon juice. Set aside for at least 2 hours, or until you are ready to serve the dish. Gently reheat the mushrooms over low heat while you grill the fish.

Prepare the grill.

Cooking

1. Rub the fish on both sides with a little olive oil and season lightly with salt and pepper.
2. Grill the fish for 5 or 6 minutes, or until done, turning it to make a cross-hatch pattern. (Halfway through the grilling time, flip the fish over to cook the other side. Make a cross hatch pattern on this side, too.)
3. Remove from the grill and keep the fish warm.
4. While the fish is grilling, melt the butter in a medium-sized saucepan.
5. Add the diced shallots, cover, and sweat the shallots until they are soft.
6. Add the haricots verts and sauté them quickly over high heat.
7. Season with salt, pepper, and lemon juice. (If you are using pousse pieds, do not add any salt.)
8. Place the grilled bass and haricots verts on warm dinner plates. Spoon the reheated mushrooms over the fish and vegetables.
9. Garnish with sprigs of chervil and lemon wedges. Serve immediately.

Tip

In this recipe, the mushrooms are marinated in their own liquid and should be allowed to sit for a few hours so the flavors have a chance to develop. If the mushrooms are large, quarter them. Otherwise, leave them whole.

Preserving Chanterelles

Drying

Thinly fleshed chanterelles such as *Cantharellus tubaeformis, Cantharellus aurora*, and *Craterellus cornucopioides* are well-suited for drying. They maintain their aroma and consistency quite well. Only flawless mushrooms that are not infested with larvae or parasites should be air-dried. Large mushrooms are split. The mushrooms can also be quick-dried using a special dryer. When the mushrooms are totally dry, they should be put in a jar with a tight-sealing lid. Mushrooms stored in this manner should remain useable for years if they are stored in a cool dark area. Prior to cooking, the dried mushrooms should be soaked in water covering them completely. As far as the species *Cantharellus tubaeformis, Cantharellus aurora*, and *Craterellus cornucopioides* are concerned, they need to soak for only 15 minutes. The dried mushroom can also be crushed into mushroom flour for use as seasoning in soups and sauces. Ordinary chanterelles are also suitable for this, but they should be sliced a few millimeters thick prior to drying.

Freezing

Cleaned and precooked (i.e., cooked in their own broth) mushrooms can be frozen. The mushrooms should be allowed to cool down quickly before they are frozen. All chanterelles, *Cantharellus aurora, Craterellus cornucopioides*, and *Hydnum repandum* are suitable for freezing. Older specimens of *Hydnum repandum* often have a somewhat bitter taste. In order to make the taste less pronounced, one can, during the precooking process,

skim off the bitter flavorings that float up with the foam to the surface. It is also possible to add water during precooking and then pour out the water. In other words, the mushrooms are slightly parboiled.

Mixing Mushrooms

Many chanterelles have a spicy flavor. They can therefore easily be mixed with mild-tasting mushrooms. A mixture of equal parts chanterelles and boletes (e.g., *Leccinum aurantiacum* or other edible *Leccinum* species) still tastes like chanterelles, but milder. In addition, one should try to avoid mixing chanterelles with, for example, *Agaricus campestris,* which grows wild and has a strong taste. The taste of such a mixture is as incompatible as if one were to use both lemon and vanilla in a cake at the same time. It is sufficient to use only one spicy mushroom in a mix at a time. The various chanterelle species, however, contain similar flavorings and can therefore be mixed with each other. Cultivated white mushrooms, the taste of which is much milder than the wild variety, can be added if one does not have enough chanterelles. However, it is better to use oyster agaricus, which is now being cultivated commercially. A good "late fall mix" of wild mushrooms can contain chanterelles, *Cantharellus tubaeformis, Hydnum repandum,* mild Russulas, and boletes. If one ends up with too much of *Hydnum repandum*, one can precook it by itself, and pour out the somewhat bitter-tasting broth. One can also skim off the "lather" that is formed on the surface and collects the bitter flavorings.

Terminology

Under "nomenclature," the so-called authority citations are explained. In this book, they have only been used on the full pages with pictures of the species. This practice may sometimes seem confusing, but it explains why the names of some species change occasionally.

Agaricon
Greek name of a mushroom already used for medical purposes in antiquity (see page 13).

Ascomycetes
Fungi that produce spores in asci (see page 11).

Basidium
Spore carrying structure of basidiomycetes (see page 11).

Boletus
The Roman name of the species Caesar's Mushroom (see page 11). Today it refers to a group of mushrooms, the boletes.

Chiastobasidia
Basidia whose nuclei divide at a right angle to the longitudinal axis (see page 12).

Cryptogamia
Linnaeus's designation of plants reproducing by means of spores (see page 10).

Cyanobacteria
Blue-green algae (see page 11).

DNA
Deoxyribonucleic acid. The cell content of an organism is called protoplasm. It consists of cytoplasm and nucleus. Many fungi have more than one nucleus. The chromosomes, containing the genetic material, are located in the nucleus. The name is derived from the Greek word *chroma*, which means color, as they can be seen (using a microscope) by means of special coloring methods. The chromosomes are made of genes, which

have been found to consist of DNA molecules. Each gene has its own place inside a chromosome. Such a place is referred to as a locus. The genes are replaceable, i.e., variant DNA molecules may be found in the same locus. A gene set determines the genotype, an organism's appearance and characteristics.

DNA analysis

By analyzing the DNA sequences in the genes, one can compare different organisms to each other and identify them. DNA sequences are as unique as fingerprints. DNA analysis is a result of the development of DNA technology.

DNA technology

Recently, scientists have developed a technique of producing large amounts of the same DNA from small amounts. One such technique is the PCR (from polymerase chain reaction) method. A chain reaction causes the number of molecules to multiply by a factor of millions in a few hours. The PCR method was awarded the Nobel Prize in chemistry in 1993. DNA analysis can now be performed on very small samples of an organism. An application of this is to determine which fungus is part of a mycorrhiza. Only a few cells from the tip of the root are necessary to determine the species.

Ectotrophic

See Mycorrhiza below.

Eukaryote

An organism with cells containing nuclei (see page 11).

Fungi

The Latin name of most mushrooms, coined by the Romans (see page 10). The mushroom kingdom.

Hymenium

The spore-producing surface of mushrooms, which carries the basidia (see page 11).

Hyphae

The threads that make up the mycelium of a fungus.

Mycorrhiza

The ectomycorrhiza is described on page 27. It is characterized by the hyphae of a mushroom forming a mantle around a tree's short root and hyphae growing between root cortex cells. Aside from this, there is endomycorrhiza. It appears among many angiosperms and is characterized by the fungal hyphae penetrating the cells, in some cases creating bushlike ramifications and nests inside the cells. The endomycorrhizal relationship is vital for most higher plants (Smith and Read 1997). The chanterelles form ectomycorrhizal relationships with many different species of trees. *Cantharellus cibarius* does this with fir, pine, hornbeam, birch, beech, and oak, to name a few.

Nomenclature

The colloquial names given to chanterelles are discussed on page 30. For as long as humans have been able to speak, they have classified the things found in their environment and named those of interest. Scientific classification and nomenclature was, however, something that did not emerge until fairly recently. Classifications were first based on detailed descriptions. A name issued was formulated in descriptive terms. Through Linnaeus, for example, genera and species were formalized. A species, for example, is a taxon, a taxonomic unit. The genus is another, of higher level; the family is the next one. Scientific naming of plants and fungi is now based on Linnaeus's work *Species plantarum*, first edition, May 1, 1753. Nomenclature (from the Latin word *nomen*, meaning name, and *clatura*, meaning to designate) is based on some basic principles. Some of these are

- The name is based on a type (type specimen).
- The name is based on priority.
- Each taxon can only have one correct name.
- All names are Latin-based.
- The rules are retroactive.

It is therefore better to talk about scientific names than about Latin names. The reason that a name is in Latin is only a result of the principles. The name of a species consists of the genus, the species epithet, and the authority citations. The latter provides some indication of the history of a species. The species epithet forms the basis for this. Let us take a few examples. The mushroom Horn of Plenty has been given the scientific name *Craterellus cornucopioides* (L: Fr) Pers. In 1753, Linnaeus referred to it as *Peziza cornucopiodes* in *Species plantarum*, while, in 1821, Fries gave it the name *Cantharellus cornucopiodes*. In 1825, Persoon included it in the genus *Craterellus*, which we still do today. This sequence of reclassification can be discerned from (L: Fr) Pers. The ordinary chanterelle has the scientific name *Cantharellus cibarius* Fr. In this case, we cannot go back to Linnaeus, since he called it *Agaricus cantharellus*, and the species currently is included in the genus *Cantharellus*. One of the rules of the nomenclature is that the genus and the species cannot have the same name, although this rule does not apply to the naming of animals. In 1821, Fries introduced the name *Cantharellus cibarius*. According to the rules of priority and retroactivity, the last-mentioned name applies. However, the priority rule may sometimes not apply. For example, names found in Fries's *Systema* are preferred over other names. An older synonymous name may then become invalid. A few additional rules apply when a species is redescribed. However, explaining that goes beyond the scope of this book.

pH
A unit of acidity level (see page 28).

Procaryota
Organisms that lack cell nuclei (i.e., bacteria).

Stichobasidia
Basidia in which the cells are divided lengthwise in relation to the basidium (see page 12).

Botanists and Mycologists in History

Batsch, August Johann Georg Karl
German botanist and mycologist, 1761–1802.

Bock, Jerome (Hieronymus Tragus)
German botanist and physician, 1498–1554. Author of one of the first modern herbal books. (herbals)

Bresadola, Giacomo
Italian priest and mycologist, 1847–1929, active in Trent.

Bulliard, Jean Baptiste François
French botanist and mycologist, 1752–1793.

Clusius, Carolus (Charles de l'Escluse)
Dutch botanist, 1526–1609. Traveled in Portugal and Spain and worked later in Vienna and finally as a professor in Leiden, 1593–1609 (see pages 14, 16).

Desmazière, Jean Baptiste
French botanist, 1786–1862.

Dillenius, Johann Jakob
German physician and botanist, 1684–1747. His works laid the foundation for classifying fungi based on the appearance of the hymenium. They formed the basis for the work of later botanists and mycologists (see page 34).

Dioscorides, Pedanios
Greek physician, circa 30–70, worked under Claudius and Nero. He was the author of antiquity's most known pharmacological book, *De materia medica*, a pharmacological standard work until the Renaissance (see page 13).

Fries, Elias Magnus
Swedish botanist and mycologist, 1794–1878. Fries received the degree Master of Science in 1814, in Lund, Sweden, and became adjunct (assistant) professor in 1819. In 1828, he was appointed demonstrator in botany. He moved to Uppsala in 1834 and first received a professorship in Applied Economics. In 1851, he became professor and head of the botanical museum. Fries's extensive scientific production included many works that

form the foundation of modern mycology (see pages 22, 23). He has also been referred to as the Linnaeus of mycology. (For more about Fries, see pages 15, 22.)

Juel, Hans Oscar
Swedish botanist, 1863–1931.

Karsten, Petter Adolf
Finnish mycologist, 1834–1917. He published a number of books and articles about Finnish fungi, e.g., *Mycologia fennica*, 1871–1878, and *Rysslands, Finlands och den skandinaviska halföns hattsvampar* (Cap Fungi in Russia, Finland and the Scandinavian Peninsula), 1879–1882.

Krombholz, Julius Vincenz Edler von
Czech physician and mycologist, 1782–1843; professor of pharmacology in Prague (see pages 15, 20).

Lindblad, Matts Adolf
Swedish botanist, 1821–1899; assistant professor in Uppsala, 1855–1873; student and friend of Elias Fries.

Linnaeus, Carolus (Carl von Linné)
Swedish physician and botanist, 1707–1778. He studied medicine in Leiden (Holland) and in Uppsala, where, in 1741, he received a professorship in botany and anatomy. Linnaeus had a great impact on mycology by creating a formalized nomenclature in the field of biology. In his works *Systema naturae* (1735) and *Species plantarum* (1753) he laid the foundations for a systematic nomenclature in both zoology and botany. Linnaeus was also interested in investigating how to best preserve Sweden's natural resources. These efforts produced such writings as *Flora oeconomica* and accounts of his travels to different areas of Sweden (see pages 34, 88–89, 105).

Maire, René
French botanist, 1878–1949.

Martial (Marcus Valerius Martialis)
Roman poet, 40–102 AD; described the life of the upper classes in Rome (see page 88).

Mattioli, Pietro Andrea
Italian physician, 1500–1577; provided comments for Dioscorides's works.

Persoon, Christiaan Hendrik
German–Dutch botanist and mycologist, 1761–1836; worked in Göttingen, Germany, and Paris, France. One of the pioneers of modern mycology,

whose works still form the foundation of mycological research (see pages 15, 18, 22).

Pliny (Gaius Plinius Secundus)
Roman natural scientist, 23–79 AD; perished during the eruption of Vesuvius in 79 AD. His work *Naturalis historia libri XXXVII* (37 volumes) contains information about fungi (see page 11).

Post, Hampus Adolf von
Swedish geologist and chemist, 1822–1911; collaborated also with Elias Fries in creating the archive for mushroom plates that were collected for the Royal Swedish Academy of Science (see page 22).

Quélet, Lucien
French physician and mycologist, 1832–1899.

Romell, Lars
Swedish mycologist, 1854–1927. One of the foremost mycological researchers, who continued in the tradition of Fries.

Schaeffer, Jacob Christian
German priest and mycologist, 1718–1790. His most known work is *Fungorum qui in Bavaria et Palatinatu circa Ratisbonam nascuntur icones*, published in 1762–1774 in four volumes with color plates (see pages 14, 18).

Steerbeck, Frans van
Flemish priest, 1630–1693. His most important contribution to mycology was his work *Theatrum fungorum*, the illustrations of which are based on Clusius's *Codex*, which was acquired by Steerbeck.

Vaillant, Sébastien
French botanist, 1669–1722. His mycological work *Botanicon parisiense*, which contained copperplate engravings by the artist Claud Aubriet, was published in 1727, see page 14.

Index

H

J

K

L

M

N

O

P

Q

R

S

T

V

W